Irrigation in the Bajío Region of Colonial Mexico

Dellplain Latin American Studies

Irrigation in the Bajío Region of Colonial Mexico

Michael E. Murphy

This book provides detailed histories of colonial water systems in four localities in the Mexican Bajío—Celaya, Salvatierra, Valle de Santiago, and Querétaro. It includes studies of irrigated agriculture, hydraulic technology, and water law in the region.

The local histories richly illustrate, through the patterns of irrigation, the interactions between the social and economic power of the landowning elite and the interventions of the authoritarian colonial bureaucracy. The study of hydraulic technology reveals colonial Mexico as being in only partial contact with the centers of innovation in Europe but possessing a local tradition of excellent architecture and some competence in surveying that found occasional expression in water works. An analysis of water law shows a system of engrafted rules regarding possession, easements, and natural water courses taken from Castillian law.

Michael E. Murphy has a Ph.D. in geography from the University of California at Berkeley and is currently a free lance writer.

DELLPLAIN LATIN AMERICAN STUDIES

PUBLISHED IN COOPERATION
WITH THE DEPARTMENT OF GEOGRAPHY
SYRACUSE UNIVERSITY

Irrigation in the Bajío Region of Colonial Mexico

Michael E. Murphy

Dellplain Latin American Studies, No. 19

LONDON AND NEW YORK

First published 1986 by Westview Press, Inc.

Published 2018 by Routledge
52 Vanderbilt Avenue, New York, NY 10017
2 Park Square, Milton Park, Abingdon, Oxon OX14 4RN

Routledge is an imprint of the Taylor & Francis Group, an informa business

Library of Congress Catalog Card Number: 86-50947
ISBN 13: 978-0-367-01127-7 (hbk)
ISBN 13:978-0-367-16114-9 (pbk)

Contents

Figures

Acknowledgments

I began this project after ten years in the practice of law with a record of legal scholarship but with no previous experience in Mexican archives. Under these circumstances, I did not tire in my search for advice and assistance. I owe my first debt to Professors James Parsons and Bernard Nietschmann of the geography faculty at the University of California, Berkeley, for their encouragement of my research, and to Professor Woodrow Borah, who wrote a number of personal letters of introduction to Mexican scholars. I was privileged to have productive discussions with Wigberto Jiménez Moreno, Alejandra Moreno Toscano, Antonio Pompa y Pompa, Jan Bazant, Andres Lira, Fr. José de Jesus Orozco, and Fr. Lino Gómez Canedo. I owe a special debt to Roberto Moreno, who wrote an eloquent letter of introduction that greatly facilitated my work in Guanajuato and Querétaro. Two other *norteamericanos* with similar research interests worked at the AGN for periods during my stay. Michael Meyer stimulated my thinking on water law, and John Super guided me to new archival sources on Querétaro. I preserve a warm memory of many Mexicans whom I met in my search for materials in the Bajío: Sr. Enrique Jiménez Jaime, the chronicler of Celaya; Fr. Jesus Guzmán, who spent many hours in the Franciscan archive to assist me in my work; Ing. Francisco Parra Acosta, who put me in contact with two elderly *ejidatarios*; Dr. Mariano González Leal, who first suggested that I research the records of *capellanías* in Morelia; Dr. Benjamin Lara, who introduced me to the remarkable farmer-scholar, Basilio Rojas; and Sr. Rojas, who allowed me to copy some useful documents in his private archive. But I owe my most profound debt to the chronicler of Salvatierra, Vicente Ruiz Arias, who generously shared with me his precise and encyclopedic knowledge of the colonial history of Guanajuato. Sr. Ruiz referred me to vital sources that I could not otherwise have found and pointed out unsuspected relationships that enabled me to assemble the scrambled mosaic of sources on Salvatierra and the Valle de Santiago into a coherent pattern. If I had not received the benefit of his detailed mastery of archival materials, I could not have successfully completed the chapters on these two areas.

Abbreviations

Archival

AA	Augustinian Archive, Mexico City
ACM	Archivo Casa Morelos, Morelia
AFC-AP	Archivo Franciscano de Celaya, Archivo Provincial
AFC-CC	Archivo Franciscano de Celaya, Convento de Celaya
AFC-CSC	Archivo Franciscano de Celaya, Convento de Santa Clara
AGN	Archivo General de la Nación
AGN-A	AGN, Ayuntamientos
AGN-C	AGN, Civil
AGN-G	AGN, General de Parte
AGN-H	AGN, Historia
AGN-M	AGN, Mercedes
AGN-P	AGN, Padrones
AGN-R	AGN, Reales Cédulas Duplicadas
AGN-T	AGN, Tierras
AHML	Archivo Histórico Municipal de Leon
AMS	Archivo Municipal de Salvatierra
BN-AF	Biblioteca Nacional, Archivo Franciscano
CAAM	Cathedral Archive Archbishopric of Mexico
C-AD	Condumex, Adquisiciones Diversas
LADC	Latin American Documents Collection, University of Texas at Austin
MNA.AH	Museo Nacional de Antropología, Archivo Histórico
MNA.AM-ACM	Museo Nacional de Antropología, Archivo de Micropelículas, Archivo Casa Morelos
MNA.AM-Qro	Museo Nacional de Antropología, Archivo de Micropelículas. Archivo Notarial de Querétaro

Legal and Governmental

NR	*Nueva Recopilación*
NovR	*Novísima Recopilación*
RLI	*Recopilación de Leyes de las Reynas de Indias*
SP	*Siete Partidas*
SRH	*Secretaría de Recursos Hidráulicos*

Introduction

> There can be no rich wheat harvests unless water is diverted from rivers and conducted a long distance through irrigation canals. This system of irrigation is most notable in the beautiful plains along the river Santiago, called the Río Grande, and in the plains between Salamanca, Irapuato and León.
>
> Alexander von Humboldt[1]

In his *Political Essays on New Spain*, Alexander von Humboldt called attention to the importance of irrigation in the wheat production of colonial Mexico. He cited, in particular, the irrigation systems in the plains that form the western part of an area known as the Bajío, a region of interconnected valleys in the present states of Guanajuato and Querétaro (Figure 1). The practice of irrigation in the region had a long history that began with the first agricultural haciendas and communities of Spanish and Indian settlers. Extensive systems of irrigation existed not only in the western Bajío described by Humboldt, but along the River Laja and certain smaller streams toward the east. In the environs of Querétaro, an elaborate system of water distribution supplied a large urban population as well as a complex of gardens and wheat-growing haciendas.

The colonial water systems of the Mexican Bajío will be the subject of this monograph. The focus is a narrow one, but the human use of water is linked to many facets of social and economic life: a study of water use is inevitably a study of broader aspects of culture. It is a rewarding study quite independently of the importance of water systems in the larger context of regional or national history. Whether a centrally important or a marginal activity, the exploitation of this vital resource, if properly understood, can present something of a microcosm of past cultures.

In contrast to the attention accorded to pre-Hispanic irrigation systems,[2] very little has been written on the subject of water management in the colonial era. There are a few essays on irrigation communities on the northern frontier;[3] Prem has studied in detail the small Cotzala valley in Puebla; Chávez Orozco has written a brief comment on the

---

[1]Humboldt, III, Book 4, Chap. 9, p. 75.

[2]See articles cited in Martínez Ríos, pp. 13-18.

[3]Arneson; Glick (1972); Hutchins; Ostrom, pp. 27-40; Ressler; Simons.

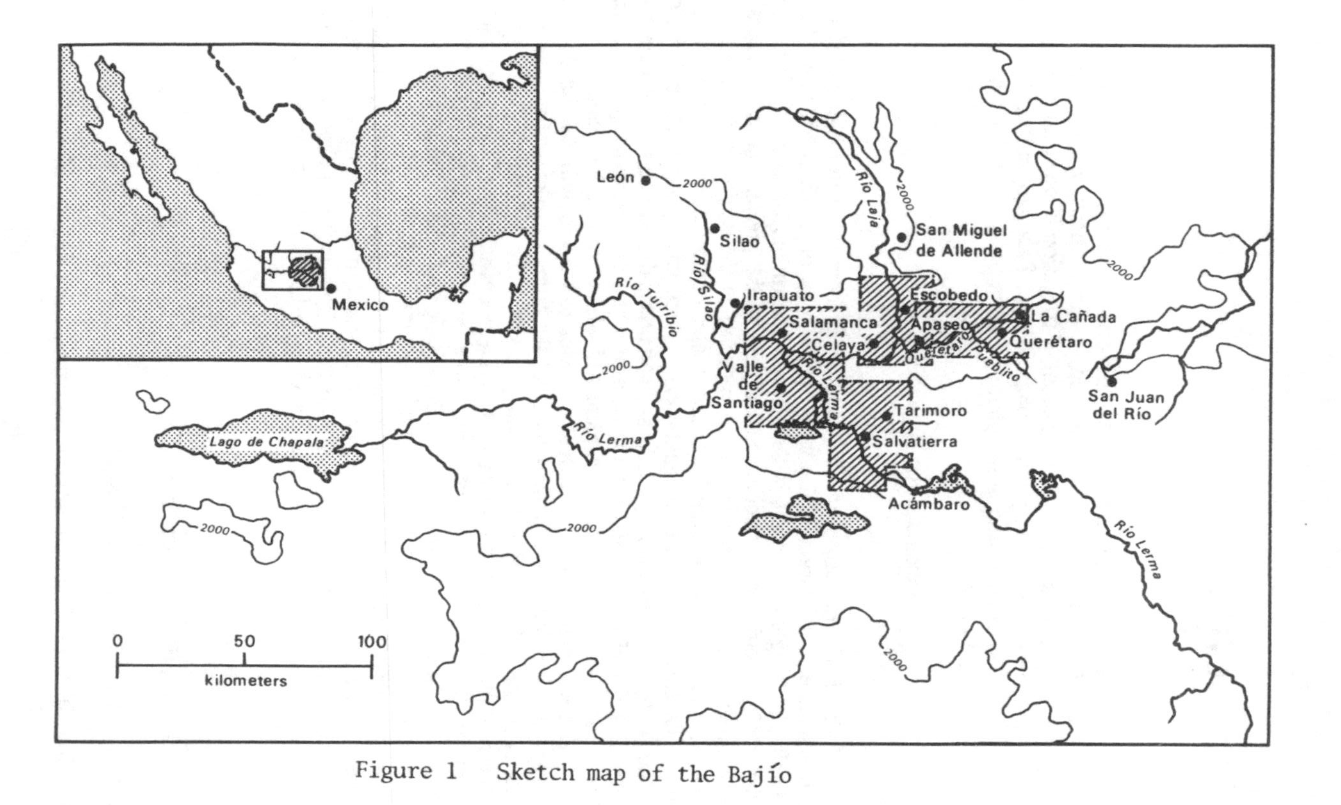

Figure 1 Sketch map of the Bajío

general topic; Taylor has contributed an essay on water law;[4] and certain other works touch incidentally on the subject.[5] The literature on other parts of Latin America is equally sparse.[6] Yet one frequently encounters assumptions as to the importance of water use in the economic history and social organization of the colonial period. A study of the Bajío region, with its long history of relatively complex water systems, offers an appropriate way to begin to test these assumptions.

The first part of this study consists of a detailed historical reconstruction of irrigation and urban supply systems in Celaya, Salvatierra, Salamanca, and Querétaro. The remainder consists of chapters on irrigated agriculture, water law, and the technology of water works. The four local studies, besides advancing certain independent inquiries, will lay the groundwork for the later chapters. In particular, they will attempt to establish a concrete context for the discussion of law and technology so that these important aspects of culture can be viewed, not in the abstract, but as part of a cultural landscape. The focus of the later chapters, however, will extend beyond the local studies to include all areas of the Bajío and, when appropriate, other regions of Central Mexico.

In addition to the themes discussed in the later chapters, the local studies will pursue three principal lines of inquiry. The first pertains to the issue of water use and social organization: was water management a significant factor in social cooperation? Conversely, did colonial institutions facilitate or impede the cooperative exploitation of water resources? The second inquiry concerns water and land tenure: was there a struggle for the control of water rights as distinct from land? And did the practice of irrigation itself affect patterns of land ownership? Finally, as a contribution to economic history, the local studies will seek to establish a chronology for the establishment of water systems and will assemble data on the investment of labor and capital in water works. The chapters on agriculture and technology have, of course, further relevance to economic history.

The Bajío has no clearly defined physical boundaries. The core of the region, lying between Querétaro, Salvatierra, and León, is surrounded by peripheral areas that, in conventional usage, may or may not be included within it. Under the broadest definition, the region is bounded by the Sierra de Guanajuato to the north, the Sierra Gorda and the Sierra de Agustinos to the east, the neo-volcanic axis to the south, and Los Altos of Jalisco to the west. As so defined, the Bajío extends roughly 200 kilometers from east to west and embraces the peripheral areas of San Juan del Río and Acámbaro, both included in this study. All of the region, except the San Juan del Río area, is drained by the River Lerma and its tributaries. The land shares a common topography and certain elements of a common history.

The flat, interconnected valleys of the Bajío, lying between 1,700 and 2,000 meters above sea level, are old lakebeds formed by the disruption of drainage patterns through volcanic activity beginning in the Cretaceous period. According to Waitz, the lakes probably reached their greatest extent in recent geological times during the Pleistocene pluvial period. The valleys are underlain by alluvial and lacustrine sediments that extend to a depth of 300 meters or more.[7] The region is scattered with springs and small swampy areas, and the water table commonly lies ten meters or less below the

---

[4]Prem; Chávez Orozco; Taylor (1975).

[5]Rojas R., pp. 85-104; Sandoval, pp. 134-46. Also Gómez Pérez; Luquín.

[6]Borde and Góngora; Chincilla Aguilar; Fernández y Simon; Keith; Sherbondy; Villanueva U. and Sherbondy.

[7]Waitz, passim.

surface.[8] Most springs are of only local importance, but the relatively copious and perennial springs on the valley floor near Apaseo and in the ravine east of Querétaro, known as the Cañada, have played a very important role in the settlement history.

Located within the southern range of the subtropical calms, the Bajío has a moderately arid climate with a markedly seasonal pattern of rainfall. The annual precipitation of about 650 mm falls mostly in three summer months in the form of afternoon thundershowers. A high evapotransporation rate, caused by the pattern of relative humidity, temperature, and insolation produces a degree of aridity not found in temperate zones with a similar rainfall. The region nearly always suffers a period of water stress in the spring, and the failure of summer rains occasionally brings severe droughts. The watersheds surrounding the Bajío do not support perennial streams of much importance. The only large perennial stream, the Lerma, arises in the relatively humid valley of Toluca. The porous basalt of the mountains surrounding the valley, still largely covered with forests, and the lakes within the valley provide the reservoirs that feed the upper tributaries of the river during the dry season.[9] After it enters the Bajío at Acámbaro, the river flows northwest past Salvatierra and emerges into the heart of the Bajío near Salamanca. Here the river turns back to the south, forming a large "U" that encloses an expanse of rich bottom land known as the Valle de Santiago. The flow of the river diminishes greatly in March and April, but in the colonial period it was still sufficient to power flour mills in Salvatierra throughout the year.[10] Two topographical features have influenced the exploitation of the river. Before entering the broad valleys north of Salvatierra, the river passes over a small waterfall and descends rapidly enough in elevation that water can be conducted easily by canals to the land lying on either bank to the north. Not far downstream the river splits into several channels, including two major branches, that rejoin near Salamanca. This division of the river facilitated the early construction of dams and canals.

The River Laja originates in a basin about 4,800 km^2 north of San Miguel Allende with an average annual rainfall of only 500 mm. Although much of the land within this drainage basin was once covered with oak woodland, the vegetation is now severely degraded, and the soil is thin and eroded to varying degrees.[11] During the heavy summer rains, the runoff is funneled southward to the Lerma in torrents that occasionally exceed the latter in volume but seldom last more than twenty-four to seventy-two hours. The volume of flow varies directly and immediately with the amount of rainfall. When the river emerges into the main expanse of the Bajío north of Celaya, it has the characteristics of a graded stream, forming a trench roughly eight meters deep and fifty meters wide that curves around Celaya and proceeds west to join the Lerma. The Laja is, however, fed by a few springs that provide a modest perennial flow. The average monthly flow at a point south of San Miguel Allende during a twenty-six year period varied from 39.4 million m^3 in September to 2.8 million m^3 in March.[12]

---

[8]Mexico, SRH (1975), section 7.7. A colonial canal near Celaya actually tapped groundwater. AGN-T, Vol. 1362, Cuad. 5, f. 5; Vol. 1390, Exp 3.

[9]Tamayo, II, pp. 152, 399.

[10]Florescano and Gil Sánchez , p. 64. For the hydrology in modern times, greatly affected by upstream dams and canals, see Mexico, SRH (1970), III, p. 11-00.1.01-13.

[11]Mexico, SRH (1973), passim.

[12]Martínez Luna, p. 74.

The hills surrounding the plains of the Bajío form a ring of other small drainage basins with steep topography, little vegetation, and thin soil. During the summer rains, the water runs off quickly into a series of arroyos and small streams. West of Querétaro, a characteristic stream, the river of the Pueblito, drains a dry, unforested basin of about 500 km^2 before joining the small creek that flows from the springs in the Cañada; together they form the River Querétaro. In a recent five year period, eighty-three percent of the annual flow of the River Pueblito occurred in the months of June through September and only 2.6 percent in the months of January through April; during the rainiest month, August, the flow was 7.1 million m^3. In the northwestern Bajío, other tributaries of the Lerma form two principal river systems, the River Silao and the River Turbio, both intermediate in size between the Laja and the Pueblito and with similar patterns of flow.[13] In the valley of San Juan del Río, a complex of seasonal streams forms two principal subsystems, the River San Juan and the River Caracol, that converge on the valley to form the River Tequisquiapan and flow into the Panuco river system to the northeast.

Our study of the water systems of Celaya, Salvatierra, Salamanca, and Querétaro will reveal a varied local history that took place within a certain common historical context. Before the conquest, the Bajío lay north of the region of settled agricultural societies. It was settled by Spaniards in the latter half of the sixteenth century in the face of vigorous opposition by nomadic Indians. Cattle ranches and towns were the principal forms for settlement. A map of the northeastern Bajío in 1580, reproduced by Jiménez Moreno, depicts valleys populated by vast herds of cattle and hills filled with Indians who prey on the cattle and direct their bows and arrows at the Spanish settlers, one of whom stands transfixed with an arrow.[14] With official encouragement, many Spanish settlers grouped in a series of towns that served both as military bases in the campaign to pacify the Indian tribes and as centers of agricultural production to supply the mines to the north. The construction of irrigation systems accompanied the founding of Celaya and Salamanca.

The early Spanish colonization of the Bajío attracted a substantial migration of Tarascan and Otomí Indians from the south and east. Although many of these immigrants settled in villages founded by civil and religious authorities, the Indian pueblos of the Bajío were fewer, more widely scattered, and generally less well endowed with land than Indian communities in the areas of older settlement. Most of the Indian population lived in the towns or on Spanish haciendas;[15] a census of the *intendencia* of Guanajuato in 1793 records only fifty-two Indian pueblos.[16] A pueblo at the eastern border of the Bajío, Querétaro, had an altogether unique history. Founded by an Indian leader who allied himself with the Spaniards, it grew by degrees to become a major urban center, dominated by Spaniards but with traces of its Indian origins.

By the end of the colonial era, the frontier society of the Bajío had evolved into a relatively populous region with a balanced economy based on agriculture, stock raising, mining, industry, and trade. The Bajío participated in the remarkable population growth experienced by western Mexico in the eighteenth century. According to Morin, the population of the diocese of Michoacán, which included the Bajío, increased almost

---

[13]Mexico, SRH (1970). See sections regarding Rivers Pueblito, Guanajuato, Silao, Gómez, and Turbio.

[14]Jiménez Moreno, p. 90; Stanislawski, p. 46 (map).

[15]Wolf, pp. 190-91.

[16]Brading (1978), p. 19.

fivefold in the eighteenth century.[17] Cook and Borah have documented an even more rapid rate of population increase for adjacent regions of west central Mexico.[18] The population growth within the Bajío itself seems to have been moderate. If the 1793 census of the intendency is compared with an ecclesiastical census of the same territory in 1760, it shows an increase of forty-five percent in the thirty-two year period.[19] But the population density of the intendency of Guanajuato was still much higher than that of any other intendency in New Spain at the end of the colonial period.[20]

Agricultural and stock raising production increased with the population. The tithe revenues for the archbishopric of Michoacán rose from an average of 103,196 pesos for the period of 1701-1705 to an average of 369,801 pesos for the period 1796-1800. Since prices were not subject to any general inflation until the last quarter of the century, these revenues provide an indication of actual production.[21] The expansion of the agrarian economy served the needs of a complex society; according to an estimate of the *intendente* of Guanajuato, only 48.7 percent of the workforce in 1793 were engaged in agriculture; 34.1 percent were employed as miners, artisans, industrial workers, and tradesmen; and the remainder were day laborers (*jornaleros*) not tied to a particular line of work.[22]

From the beginning of the period of colonization, the haciendas of the Bajío supplied northern mines as distant as Zacatecas, but in the eighteenth century Guanajuato, at the edge of the Bajío, emerged as the most important focus of commercial relations. Mining production in New Spain increased about sevenfold in the century,[23] and Guanajuato accounted for between one-quarter and one-fifth of the coin minted in New Spain in the 1790s. The city and its environs then had a population of about 55,000, and the mines and refining operations employed no fewer than 14,000 mules.[24] In 1779, the city consumed twice as much maize as Mexico City.[25] The population of Guanajuato, like that of other cities, also constituted a major market for certain Spanish foods, above all wheat flour. Although wheat was more expensive than the indigenous maize, wheat bread was desired by a large segment of urban residents.[26]

For a pre-industrial society, the Bajío contained a very large urban population that was distributed in a network of large and small towns. Brading has estimated that as much as one-quarter or one-third of the population of the central Bajío, including

---

[17]Morin, pp. 39-83.

[18]Cook and Borah, I, pp. 300-75. Also Serrera Contreras, pp. 11-17.

[19]Morin, p. 44.

[20]Florescano and Gil Sánchez, pp. 146-48.

[21]Morin, pp. 103, 193.

[22]Brading (1978), p. 20.

[23]Brading (1971), p. 131.

[24]Brading (1978), p. 18; Ward, I, p. 43.

[25]Morin, p. 142.

[26]Brading (1971), pp. 146-69, 224-30, 248-54; Morin, p. 146 (see quantity of wheat purchased in León).

Querétaro, lived in towns of over 5,000 population.[27] West of Querétaro lay the small cities of Celaya, San Miguel el Grande, and León, as well as lesser towns such as Irapuato, Silao, Salamanca, and Acámbaro. All the towns served as points of distribution for agricultural products; some lay on important trade routes to the west and north of Mexico; and most contained textile manufacturing and artisan industries.[28] Querétaro was the center of the textile industry, specializing in woolen cloth, but other towns of the Bajío, particularly San Miguel, Celaya, Acámbaro, and León, also had small populations of weavers.[29] The towns were, however, more than economic entities; they were also religious centers and places of preferred residence for the upper class. They harbored a dense concentration of churches and convents in relation to their population and possessed substantial houses on highly valued land near the central squares.[30]

Within the agrarian economy serving these mining and urban centers, agriculture increased in importance relative to stock raising during the colonial period. In San Miguel, agricultural products accounted for only five percent of total tithe revenues in 1689 but represented sixty percent of revenues fifty years later.[31] At the end of the colonial period, the Bajío was said to be the most productive agricultural region in Mexico.[32] The fields between Salamanca and León reminded Humboldt of the "most attractive countryside in France."[33] Another acute observer, the English envoy, H.G. Ward, acknowledged that the Bajío was celebrated "as the seat of the great agricultural riches of the country."[34] The land tenure pattern of the Bajío was largely a mosaic of moderately sized haciendas and rented ranchos, but there were also some great estates and small independent farmers. Agriculture was always a commercial enterprise. Apart from the opportunities afforded by nearby markets, the weight of fiscal exactions and mortgages pressured the rural landowners to produce a substantial surplus.[35] In the eighteenth century, the *hacendados* of the Bajío enjoyed increased demand for agricultural products but faced a shortage of indigenous labor and competition from small rancheros. Many landowners chose to rent out large sections of their estates to increase revenues and secure a resident workforce. Brading has shown that much of the expansion of tillage occurred on hacienda land rented out to small rancheros. Other landowners, however, responded to the challenge by improving their land through investments--often

---

[27]Brading (1978), p. 19.

[28]Moreno Toscano, passim.

[29]Morin, p. 122. In 1793, more than 200 looms were reported in towns of the Bajío other than Querétaro, and each loom was said to employ six persons.

[30]The most cursory review of the *ramo de capellanías* in the Archivo Casa Morelos in Morelia will reveal the importance of urban real estate in the fortunes of the upper class.

[31]Galicia Morales, p. 115.

[32]Wolf, p. 188 (Joel Poinsett).

[33]Quoted in Brading (1978), p. 13.

[34]Ward, II, p. 421. Also Vol. 1, pp. 11, 44, 48.

[35]Morin, pp. 141-60.

drawing on wealth earned from commerce or mining--in hydraulic improvements, barns, fences, and land clearance.[36]

The production of maize, the staple crop, usually supplied a strictly local market, but the production of some other agricultural products, such as wheat, responded to a larger regional demand. The traveling merchant played an important role as a kind of wholesaler. Commanding several teams of mules loaded with wheat, sugar, beans, textiles, salt, soap, candles, and other goods, the typical merchant traveled a wide circuit of towns and haciendas. A portion of the wheat crop served a more specialized trade with the capital. The high quality wheat of the Bajío sold for a premium that paid for the cost of transportation to this relatively distant market. In 1770, it was reported that of the 115,000 cargas of flour consumed in Mexico City no less than 87,000 cargas came from the Bajío and other parts of the diocese of Michoacán.[37]

In modern times, the Bajío has undergone further transformations. The process of urbanization has continued. León has grown to be a city of 600,000 population with a vigorous leather-working industry, and the many small towns and hamlets have swollen in size. But Querétaro and Guanajuato have declined in relative importance. The textile industry of the Bajío was destroyed with the inflow of cheap British imports in the early nineteenth century, and the mines of Guanajuato have largely played out. Today Querétaro is an intermediate- sized city of about 180,000 population. Guanajuato, though now the capital of the state, is not much bigger than it was in the colonial period. Agriculture has retained its importance in the economy of the Bajío through the intensive cultivation of the limited expanse of good arable land. Major storage dams on the Lerma and Laja supply extensive irrigation systems, and the abundant groundwater reserves have been exploited, or over-exploited, to the point of dangerously reducing the water table in some areas. The irrigated land produces a wide range of crops under a regime of double cropping and fertilization. The Valle de Santiago and nearby alluvial land along the Lerma might be described as the agricultural showcase of the region with extraordinarily high yields of sorghum, maize, wheat, barley, and vegetables. The municipality of Celaya, a typical area, possesses 30,600 hectares of irrigated land out of a total of 51,300 hectares of arable land and produces maize, wheat, forage crops, tomatoes, and a wide variety of other vegetables.

The patterns of agriculture today present a sharp contrast with those we will study in the colonial era. For reasons we will analyze in Chapter 6, irrigation was then closely linked to the production of one crop, wheat. The local studies to which we now turn will be in large measure the stories of groups of wheat-growing haciendas, each affected by the peculiar course of local history.

---

[36]Brading (1973), pp. 228-37.

[37]Morin, pp. 141-60.

1
Celaya

The first great agricultural estate of the Bajío had its origin in 1542 when the *encomendero* of Acámbaro, Hernán Pérez de Bocanegra, brought Tarascan allies to Apaseo, an Indian town that had been founded, or at least placed, under Spanish control a few years earlier by a lieutenant of Nuño de Guzmán.[1] At the chosen site, a spring of substantial volume surfaced close to the small stream flowing westward from Querétaro (later known as the River Querétaro). The site thus presented an unusual attraction: a perennial water supply that could be easily captured for irrigation. In a compact with Indian leaders, Bocanegra conceded to the aboriginal community the right to one-half of the water of the stream below the point where it was fed by the spring; Bocanegra reserved for himself the more attractive southern bank where the spring had its origin. Within a year, dams and irrigation canals had already been built.[2]

The Bocanegra estate, expanded by land grants to family members, grew to be a large agricultural enterprise and was converted into a *mayorazgo*.[3] A description of the property in 1557 describes the farm "beside the river..., with its dwellings, barns, granaries, stock pens, threshing floors, and all the *caballerías* contained in said enclosure of Apaseo with its watercourse and irrigation canals."[4] The newly discovered mines of Guanajuato and Zacatecas gave the area importance as a source of food supplies.

In the 1560s, the agricultural frontier began to extend west to the land near the bend of the River Laja. The governor of Acámbaro, Cristobal de León,[5] and a native of Galicia, Juan de Yllanes, began to acquire and cultivate wheat land near the junction of the River Querétaro and the Laja. Other enterprising Spanish colonists were also attracted by the rich land and possibilities of irrigation in the area. In 1565 there were fifteen grants of land fronting on the river.[6] It is doubtful that many of the recipients took possession of their land; their names do not appear in the subsequent history of Celaya, and Indian hostilities may still have been a deterrent. But several references in the grants

[1]Gerhard (1965), p. 65; Jiménez Moreno, pp. 54, 66, 72.

[2]AGN-T, Vol. 187, Exp. 2, ff.63-65v.

[3]AGN-T, Vol. 675, Exp. 1, f. 75v; AGN-M, Vol. 7, f. 120 (Luis Ponce de León); Vol. 10, f. 19 (Nuño de Chávez); Vol. 8, f. 148 and Vol. 9, f. 5v (Hernando de Cordoba).

[4]Chevalier (1952), p. 155.

[5]AGN-M, Vol. 8, f. 39.

[6]AGN-M, Vol. 8, ff. 2, 29, 40v, 43, 55, 87, 100v, 109, 110v, 111, 113, 118, 190.

to "la población nueva de los labradores" suggest that the beginning of settlement had in fact been made.[7]

About 1568, several small farmers joined in a common endeavor to build an irrigation canal on the west bank of the Laja.[8] The canal was probably located near the site of the future city of Celaya.[9] Two of the contradictory accounts of this enterprise speak of four settlers who constructed the canal, and the names of four persons are indeed mentioned more often than others: Lope García, Vasco Domínguez, Pedro López, and one Arteaga (possibly Alonso Ramírez de Arteaga). The relación geográfica of 1580 very possibly refers to the farmers sharing this early canal where it states that the settlement began with a group of three or four persons. Vasco Domínguez claimed to have spent "many pesos of gold" building the canal, but it probably represented only a rudimentary beginning for later work.[10]

These small farmers lived at the fringe of the expanding agricultural estate of Juan de Yllanes. By 1571 Yllanes operated two flour mills and possessed an extensive irrigation system on the east bank of the River Laja. A canal led from the river southeast through the fields of Yllanes and Luis Ponce de León, a member of the Bocanegra family, before draining into the River Querétaro. About this time, Yllanes also appears as the owner of irrigated land on the west bank of the river. The wheat from his farms was exported to the mining centers to the north.[11] In a legal proceeding in 1571, he explained that his flour mills

> grind five or six thousand fanegas of wheat harvested in the fields that I have in this valley from which I have supplied for many years the mines of Guanajuato, Zacatecas and the Chichimec frontier, the towns of San Miguel and San Felipe and Nuestra Señora de los Lagos.[12]

Around 1570, the Viceroy Enríquez was presented with a petition to found a town in the present location of Celaya (Figure 1.1). The alcalde mayor of Guanajuato was

---

[7]AGN-M, Vol. 8, ff. 29, 43, 55, 110v. But see AGN-M, Vol. 8, f. 29, for reference to "Arteaga," a name that will recur.

[8]AGN-T, Vol. 674, Cuad. 2, ff. 1161-1172.

[9]An irrigable parcel granted to Alonso Gutiérrez in the first distribution of land is described as bordering on fields of Lope García. AGN-M, Vol. 10, ff. 41v-46. Lope García participated in the lower canal in 1576. AGN-T, Vol. 674, Cuad. 2, f. 1154. Moreover, a canal near the town was known as the "acequia de Arteaga" in the eighteenth century. AGN-T, Vol. 1362, Cuad. 3, f. 86v.

[10]None of the settlers named as builders of this early canal were given credit for the work in later awards of water rights even though the canal probably can be identified with the lower canal which later served a series of farms. Rather, a later participant in the lower canal, Miguel Muñez, was given an extra day of water in recognition of "el gasto y trabajo que tubo en la fundacion de la dicha villa y en labrar la dicha acequia grande quando al principio se comenca." AGN-T, Vol. 674, Cuad. 2, f. 1192v. See also f. 1164.

[11]Yllanes: AGN-T, Vol. 65, Exp. 5; Vol. 674, Cuad. 2, ff. 1159v-60. Ponce de León: AGN-T, Vol. 674, Cuad. 1, ff. 100-106.

[12]AGN-T, Vol. 65, Exp. 5, f. 14v.

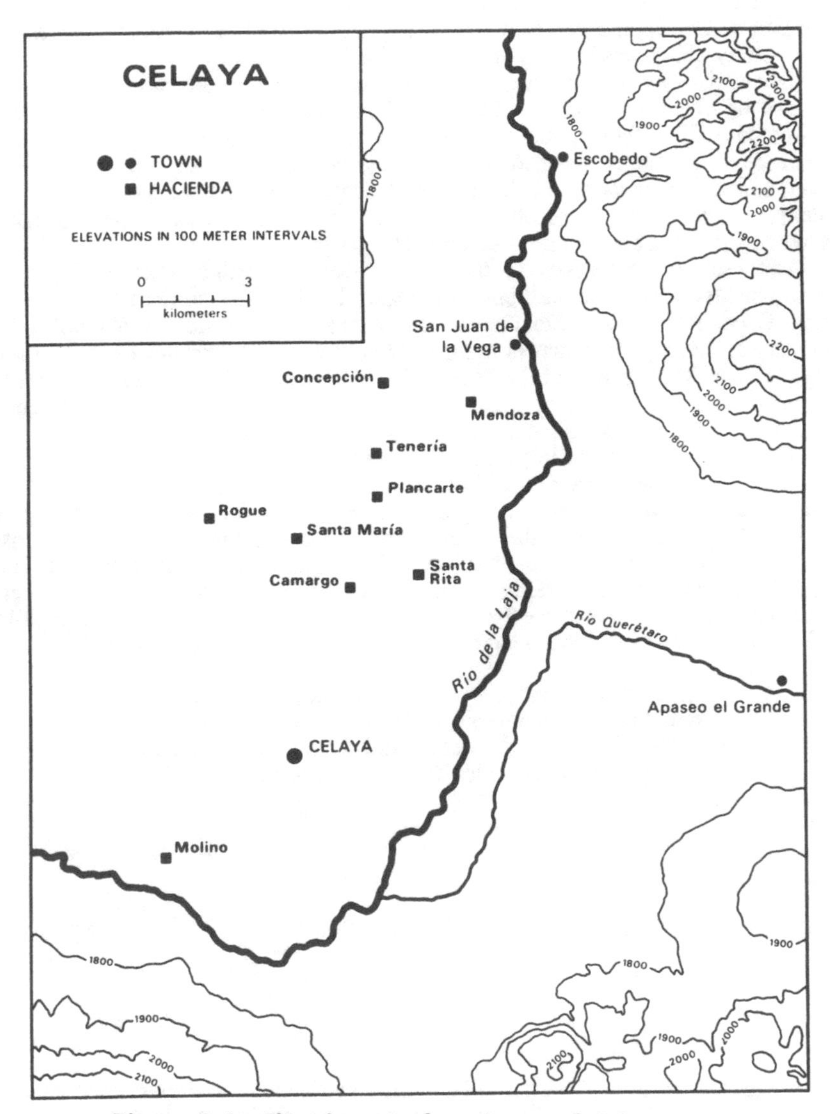

Figure 1.1 Sketch map of environs of Celaya

dispatched to investigate the proposal, and later a judicial officer, Dr. Francisco de Sandi, was sent to make further inquiries. He took the opportunity to settle a dispute between the recent settlers and Juan de Yllanes over milling charges and then, reporting back to the viceroy, recommended the establishment of a villa and the distribution of land along the west bank of the Laja.[13]

Viceroy Enríquez authorized the establishment of a villa, Nuestra Señora de la Concepción de Celaya, in a decree dated October 10, 1570.[14] This was influenced by the need to pacify the nomadic Indians of the Bajío. The site lay along the road to Guanajuato and would provide a convenient base for military operations to secure the safe transit of silver and supplies. The first alcalde mayor, Cristobal Sánchez Carbajal, was a noted Indian fighter, active in establishing block houses along the road to Guanajuato.[15] But the decree also observed that the area lay within secure territory and was well suited for supplying the "mines of Guanajuato and Zacatecas and the villas of San Felipe and San Miguel and other areas." The new villa, in fact, assumed more the character of a farming community than a military outpost.

The prospective vecinos of the villa, listed in Sandi's letter, included small farmers, landless colonists, and possibly some local cattle ranchers.[16] The letter mentions Lope García, Vasco Domínguez, and Alonso Ramírez de Arteaga, but does not refer to any of the large landowners in the area or any previous recipients of land grants. During his visit, Sandi stayed at the house of Lope García, who appears to have been a representative, if not a leading member, of the group.[17] For these settlers, the establishment of the villa signified an opportunity to acquire good title to land.

The commissioner and inspector of the region (juez de comisión y visitador), Dr. Alonso Martínez, visited Celaya in 1573 to distribute land to the vecinos of the villa. He was immediately confronted with complaints that Juan de Yllanes was diverting more than one-half (or two-thirds) of the water from the Laja to his properties and those of Ponce de León on the east bank of the river. Yllanes and a large cattle rancher, Gaspar Salvago, were also accused of taking a disproportionate share of water from the irrigation canal on the west bank. The inspector acted decisively. He confiscated three caballerías of Yllanes' land and ordered that Yllanes, Ponce de León, and Salvago be notified

> that from now on they may not take or cause to be taken...the water from said river nor irrigate their lands nor any other land with the water until the vecinos of the villa have irrigated their fields under penalty of one hundred pesos.[18]

---

[13]Velasco y Mendosa, I, p. 325; AGN-T, Vol. 65, Exp. 5.

[14]AGN-M, Vol. 38, f. 14.

[15]Powell, pp. 152-53.

[16]García had bought four caballerías of land but only cleared two for farming. AGN-T, Vol. 674, Cuad. 2, f. 1163. Diego Pérez de Lemus can probably be counted among the cattle ranchers. See AGN-T, Vol. 2691, Exp. 5.

[17]AGN-T, Vol. 674, Cuad. 2, ff. 1165-67.

[18]AGN-T, Vol. 674, Cuad. 1, ff. 100-106v.

Yllanes won at least partial relief from these actions in appeals to the audiencia. He secured orders that confirmed his right to operate a mill on the River Laja and gave him one caballería as recompense for the three he had lost.[19]

In the distribution of land, Dr. Martínez generally gave each vecino one caballería of irrigable land and one and one-half caballerías of dry land. Only seven out of thirty-two recipients failed to receive a portion of irrigable land. Most of the caballerías of irrigated land, however, were on the southeastern site of the River Laja below the confluence with the River Querétaro--an area not yet irrigated and with doubtful potential for irrigation.[20] Every vecino was required to live in the town for ten years with no unauthorized absences of more than four months on penalty of forfeiting his land.[21]

The town appears to have flourished modestly in the first decade of its existence. A church, a municipal building, and residential houses were apparently built; a presidio was founded; flour mills were established; new land was cleared; and more grants were issued. In separate accounts, the town was described as having a population of sixty vecinos in 1580 and seventy vecinos in 1582.[22]

The relación geográfica of 1580 assigns central importance to irrigation in the economy of the new settlement:

> It lies at the juncture of two rivers: one, which comes from the villa of San Miguel, runs north-south and joins another called Apaseo [or Querétaro] which runs east-southeast, and the settlers of the villa benefit from the rivers by irrigating their fields and wheat lands so that with this irrigation they are able to harvest 17 or 18 thousand fanegas of wheat; the villa is abundant in pasture for cattle and productive in fruits of Spain, peaches, quince, pomogranates, grapes, figs and others, and yields a harvest of all kinds of vegetables...Located close to the villa, there are four flour mills serving its needs; the mills are very valuable for the wheat crop which has been described...[The region] produces much wheat and much maize...[23]

There may have been an element of local boosterism in this account written by the acalde mayor. A contemporary source speaks of residents abandoning Celaya for Apaseo.[24] But the relación geográfica of Acámbaro corroborates the high level of wheat production

[19]AGN-M, Vol. 10, ff. 19v, 20v.

[20]Note grants of land in area "que solia ser ejido." Location of ejido is described in AGN-M, Vol. 38, f. 14.

[21]AGN-M, Vol. 38, f. 14.

[22]AGN-T, Vol. 65, Exp. 5, f. 9; Vol. 674, Cuad. 2, f. 1163; AGN-M, Vol. 10, ff. 16v, 17, 65v, 69v, 168v, 171v, 172, 173, 173v, 182v, 183, 194, 202v, 210v, 219; AGN-G, Vol. 2, Exp. 169; Powell, pp. 152-53. For estimates of population, see relaciones geográficas of Celaya and Querétaro in Corona Núñez, II, p. 50, and Velázquez, I, p. 29.

[23]Corona Núñez, II, p. 50.

[24]AGN-G, Vol. 2, Exp. 1297.

in Celaya (16-18,000 fanegas).[25] There can be no doubt that the settlement was firmly established with an economy based in important measure on an irrigated wheat crop.

The level of wheat production was made possible by the construction of a major irrigation system some time before 1576. The head of the canal can be located with some assurance as being precisely at the site of the dam of the Labradores of the later colonial period and of the present diversion dam of Irrigation District No. 84 near Escobedo. There are several references to the existence of two canals on the west bank of the River Laja--an "upper canal" and a "lower canal"--and one mention of a "new canal" in 1576. We know there was an irrigation canal in the area close to the town where the first land grants were made; the "new canal" and the "upper canal" must be synonymous since there is no suitable site and little subsequent history of irrigation in the river below the town.[26] Two documents help locate the exact place of origin of the canal. In 1584 one Luis de los Reyes petitioned for a license to build a flour mill "in the principal canal from which the vecinos of the villa irrigate their fields next to a small hill and a waterfall in the said canal."[27] Nine years later there was another application for a permit to build a mill near the pueblo of San Gerónimo "between it and another pueblo called San Juan about a half league from the dam of the farmers of the villa."[28] Now, to the north of the old Indian town of San Juan de la Vega where the River Laja emerges from a narrow valley, there is a peculiar dome-shaped hill and enough of a rise in the land to create a rapids in water flowing to the south. This location about eighteen kilometers to the north of Celaya, where subsequent canals have had their origin, fits well these early descriptions of the "new canal"[29] (Figure 1.2). Lying at a small elevation above the broad valley of Celaya, the site is uniquely suited for the diversion of water from the River Laja to the maximum number of farms. Ten wheat haciendas received water from the "new canal" in 1576; at the head of the canal was Pedro Hernández de los Reyes, presumably the father or a relative of Luis de los Reyes, the applicant for the milling permit.[30]

One is immediately struck by the question of how the early Spanish settlers could have built three dams on the River Laja in about ten years time (i.e., the dam of Yllanes and the dams of the upper and lower canals). The river bed today is a trench about fifty meters across and eight meters deep. It would be no trivial matter to build a dirt dam spanning the river and more difficult still to make the dam withstand summer floods.

The first vecinos did receive allocations of forced Indian labor. Shortly after the foundation of the villa, Viceroy Enríquez ordered that Indians from five towns in Michoacán be assigned to Celaya for nine months of the year.[31] The practice of exploiting forced labor continued through the century. In 1600 the Indians of Acámbaro, Apaseo, and San Pedro Chamacuero pleaded that they should be relieved of the obligation

---

[25]Corona Núñez, II, p. 63.

[26]AGN-T, Vol. 674, Cuad. 2, ff. 1147v-1154v.

[27]AGN-M, Vol. 13, f. 113v.

[28]AGN-M, Vol. 18, f. 231.

[29]The principal canal clearly flowed in this general area in the seventeenth century. AFC-CC, E-2, leg. 7.

[30]AGN-T, Vol. 674, Cuad. 2, f. 1154v.

[31]AGN-G, Vol. 1, Exp. 1304.

to contribute labor to Celaya since their pueblos had become "casi despoblado."[32] There are other records of requests for Indian labor in Celaya in 1579, 1591, and 1603.[33] But this rather ample historical record does not contain a single reference to the need for Indian labor to construct or repair irrigation dams or canals. The Enríquez order was for the expressed purpose of constructing houses for the vecinos; later documents mention the need for Indian labor in the wheat harvest and other forms of field work. While some of this labor was probably diverted to waterworks, the availability of labor for this purpose does not seem to have been a major concern of the vecinos.[34]

The settlers, it appears, built the early irrigation systems at their own expense by relatively simple means.[35] Quite likely, they made use of secondary water courses of the river, as was done in later irrigation systems of the Bajío.[36] Part of the river's flow could be diverted to a dry water course and then to a farmer's fields in a two- step process. Moreover, the river itself had a very different aspect in the sixteenth century before the effects of grazing and deforestation had been felt. The banks of the river were forested with cedar, mesquite, and willow,[37] and the stream itself had a more even seasonal flow. Dr. Sandi's report described the abundance of fish in the river--a phenomenon that has long since disappeared.[38] One can only speculate about the effect on the stream channel of the loss of vegetative cover in the drainage basin, but itis reasonable to imagine that, as in the American Southwest, the higher peak flows in modern times have deepened and enlarged the earlier channel.[39] With a smaller channel, less violent summer floods, and ready supplies of timber, the stream may have permitted the use of simply constructed diversion dams.

Construction of the canals led immediately to problems in allocating the water. The vecinos, probably of diverse social origin,[40] did not attempt to work out a solution through local institutions but rather petitioned the central government for intervention. In 1575 Viceroy Enríquez ordered a strictly egalitarian pattern of distribution:

---

[32]AGN-G, Vol. 5, Exp. 226.

[33]AGN-G, Vol. 2, Exp. 60; Vol. 4, Exp. 266 (refers to an order of April 16, 1566 (sic) to assign 100 Indians to the villa); Vol. 6, Exp. 794.

[34]One similarly finds only occasional references to forced allocations of labor for irrigation elsewhere in Mexico. Zavala (1939-46), Vol. 1-8.

[35]See AGN-T, Vol. 674, Cuad. 2, ff. 1145v-89.

[36]AGN-M, Vol. 39, f. 203v.

[37]Corona Núñez, II, p. 50; Velasco y Mendoza, I, p. 325.

[38]Ponce, I, p. 537.

[39]Dunne and Leopold, p. 706.

[40]The settlers have been described as Basques (Powell, p. 152), but their surnames do not suggest any ethnic concentration.

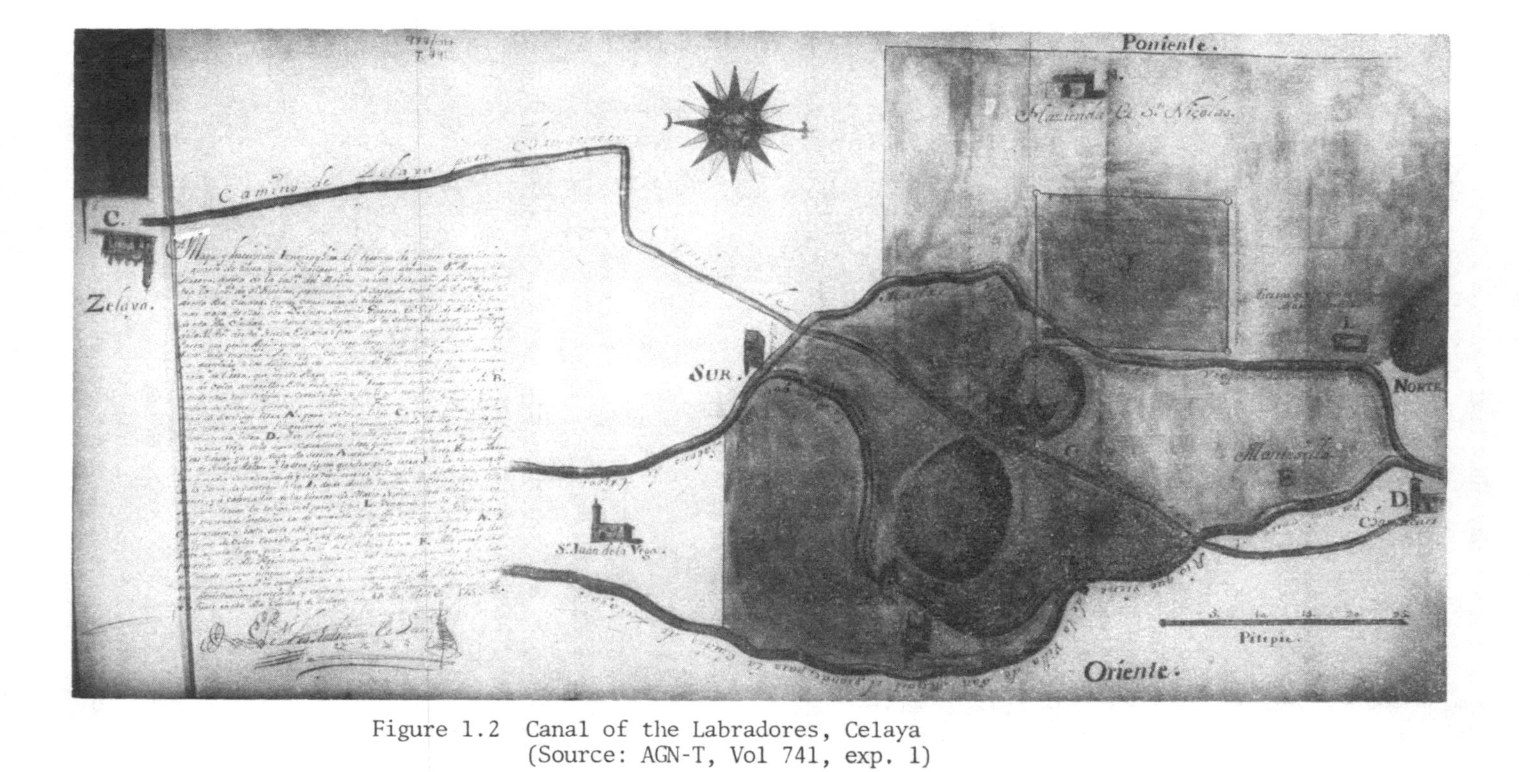

Figure 1.2 Canal of the Labradores, Celaya
(Source: AGN-T, Vol 741, exp. 1)

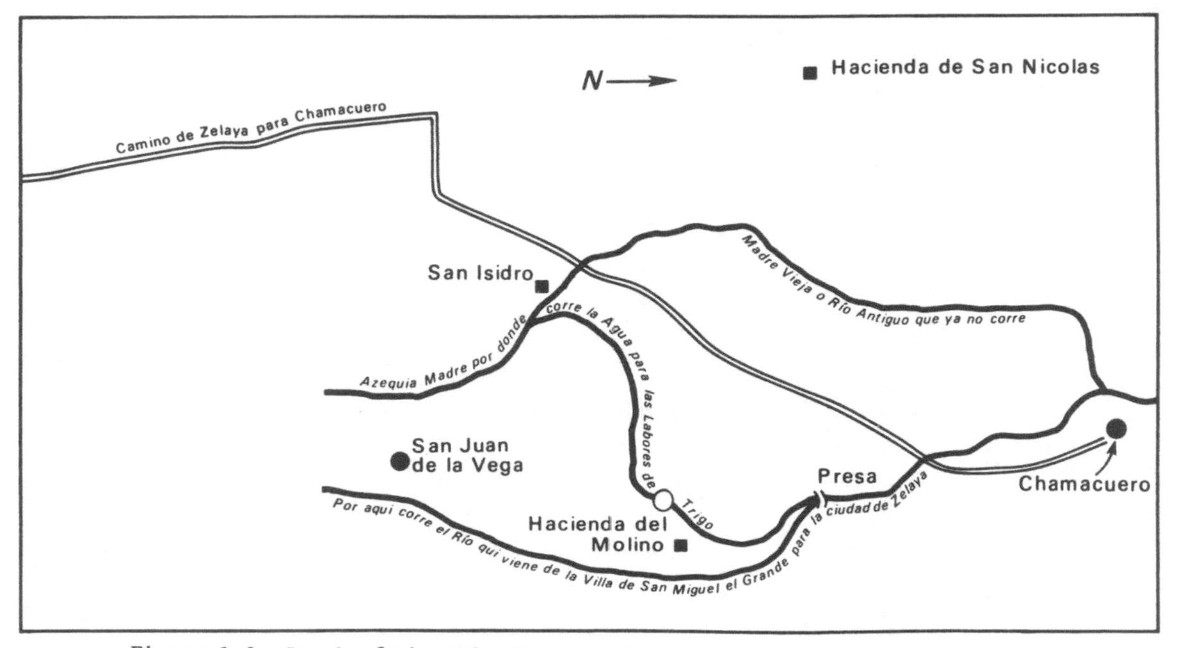

Figure 1.2 Canal of the Labradores, Celaya

This interpretation is based on the map shown opposite, dated 1753 which shows the dam of the Labradores and the upper reaches of the canal leading to the wheat haciendas of Celaya. The dam is a simple diversion type, and the canal proceeds to Celaya via the indian pueblo of San Juan de la Vega. It merged with an old course of the river Laja near San Isidro.

> Equally by shares or days, as may be appropriate, among the said fields so that one person will not receive more than another, bearing in mind the need to reserve enough water for future settlers.[41]

The alcalde mayor complied by appointing two settlers to prepare a plan of allocation. They devised a scheme whereby water would be distributed alternately through two sets of irrigation gates, "placed level and in proper order so that more water will not flow through one gate than another." Each farmer would receive a standard- sized gate except Juan de Yllanes who would receive three and one-half gates and one Juan Diaz who would receive a one-half-sized gate.

The plan met with the Viceroy's approval, but it foundered in practice. A petition to the audiencia complained that the same disorder prevailed as before. A slightly modified pattern of distribution for the remainder of 1576 was ordered by none other than Juan de Yllanes, who by this date had become a vecino of Celaya and held the offices of *alcalde ordinario* and *teniente de alcalde mayor*. The *cabildo* later approved an entirely different plan of allocation: each farmer on the two canals would be given successively the entire flow for two days, or, in the case of Yllanes, for five days. This plan was also quickly approved by the Viceroy.

The principle that each vecino would be given two days of water gained general acceptance, but the precise allocation remained unsettled. Some of the first settlers had sold their water rights in contravention of the terms of their grant. Others sought concessions in recognition of their private expense in building the irrigation systems. Moreover, there were bitter disputes over whether the practice of irrigation before the foundation of the town conferred a right to water. The cabildo took the narrow view that only a grant of irrigable land conferred water rights and denied three early settlers, including Lope García, access to irrigation from the river Laja.

In 1577, Dr. Hernando de Roble, alcalde of the royal audiencia, visited Celaya to make a definitive allocation of water rights. He made concessions to certain vecinos who had practiced irrigation before the foundation of the villa or had made personal investments in the irrigation works. Juan de Yllanes was given eight days of water--a prominent, but far from dominant, share of the available supply. Another vecino was given four days, and four vecinos were given three days. But Roble's allocation still had a roughly egalitarian character. After making a limited number of special concessions, he affirmed the principle that each settler should be given one caballería of irrigable land with the right to two days of water. Past conveyances of water rights were recognized, but future conveyances were forbidden.[42]

Roble's order was the only general adjudication of water rights of which we have record in the history of Celaya. All water rights were evidently derived either from this early allocation or from a limited number of subsequent grants of water rights. But the order was silent on the manner in which the right to a day of water was to be translated into practice. There was no plan for the rotation of irrigation turns, the location of tributary canals, the construction of irrigation boxes and gates, the sharing of costs of maintenance, or other factors that would have a vital bearing on the actual benefit conferred by irrigation. These administrative matters were left to be decided by a water judge (*executor e repartidor del dicha agua*). Roble appointed Miguel Juan, who did not himself have property on the River Laja, to police irrigation along that river. Juan would be paid a substantial annual salary of 100 pesos, which would be assessed among the

---

[41]AGN-T, Vol. 674, Cuad. 2, f. 1139v.

[42]Ibid, ff. 1139-1193.

water users in proportion to their water rights. The settlers were enjoined to comply with his orders.

> They are to carry out and obey what Miguel Juan orders with respect to the distribution of water in conformity with the rules decreed by the alcalde mayor and His Majesty under penalty of one hundred peso fine.[43]

The municipal office of water judge (juez de aguas) was authorized for all colonial towns of the Indies by royal decrees promulgated in the early sixteenth century, and it is occasionally encountered in widely scattered locations.[44] But the office seems to have had a short life within the municipal government of Celaya. The last reference to the water judge as a municipal officer occurs in 1593.[45] At some date thereafter the problems of administration and dispute adjustment were assumed by the farmers themselves, who from the beginning paid the cost of the water judge's salary.

Under the terms of Roble's order, the vecinos could be given rights to water either from the River Laja or the River Querétaro. A small irrigation canal was in fact soon built from the River Querétaro to the swampy area near the southern bank of the River Laja that had first been reserved as an ejido and later granted to the town.[46] A system of irrigation turns was in effect throughout the century,[47] but the vecinos here had to compete with the Bocanegra estate in Apaseo for water; in 1581 they complained to the alcalde mayor of Celaya that this estate was usurping an excessive amount of water from the river. The ensuing legal struggle lasted twenty-seven years and ended in the victory of the Bocanegra heirs.[48] There is no record of further irrigation on the lower reaches of the River Querétaro near Celaya. The later colonial dams in Apaseo consumed all the flow of the river during the dry season.[49] Very likely, the volume of the river itself suffered from the effects of deforestation and overgrazing. By the late eighteenth century, it would be depicted by Francisco Eduardo Tresguerras, always an accurate observer, as an insignificant arroyo (Figure 1.3).

The simple pattern of landownership that existed in the small settlement of Celaya at the time of Roble's visit would acquire new complexity in the course of the sixteenth and seventeenth centuries. The cabildo continued to grant vecindades to new vecinos. There are records of twenty-eight of these grants in the remaining years of the sixteenth

---

[43]Ibid, f. 1194v.

[44]RLI, 3:2:63; Cavo, p. 127; Bancroft Library, University of California, P-E 51:2 (El Paso, 1754-62).

[45]AGN-M, Vol. 18, f. 268. See also AGN-T, Vol. 674, Cuad. 1, f. 83.

[46]AGN-M, Vol. 10, f. 3; Vol. 38, f. 14; AGN-T, Vol. 674, Cuad. 2, f. 1146.

[47]AGN-M, Vol. 14, f. 165v; Vol. 18, f. 268; AGN-T, Vol. 674, Cuad. 2, ff. 1218, 1306, 1464.

[48]AGN-T, Vol. 674, Cuad. 1, e.g., ff. 49-79, 90, 109-111v, Cuad. 2, e.g., ff. 1092-1275, 1356-1360.

[49]AGN-T, Vol. 383, Exp. 4, ff. 71, 92v, 114v; Vol. 674, Cuad. 1, ff. 127v-29; AGN-M, Vol. 38, f. 8v; Vol. 49, f. 155v.

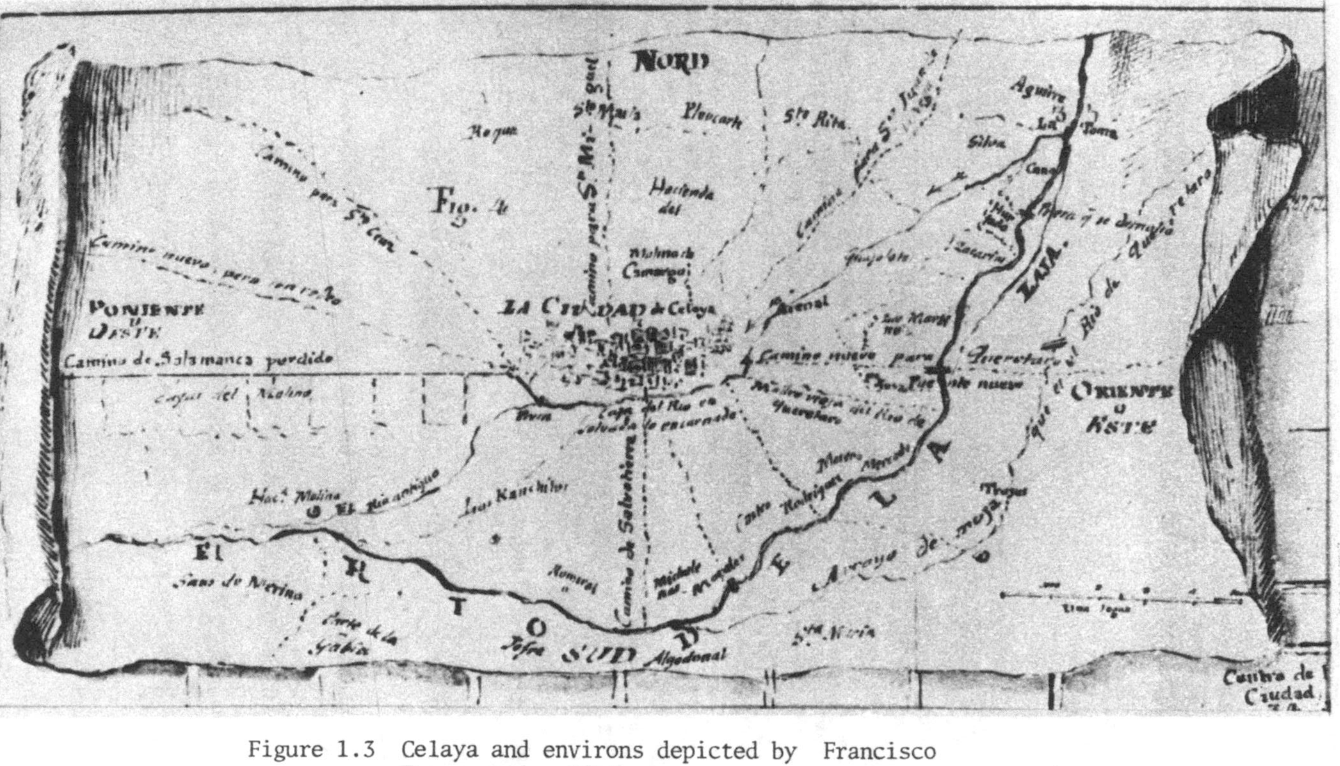

Figure 1.3 Celaya and environs depicted by Francisco Tresguerras, 1810
(Source: AGN-T, Vol 2072, exp. 1)

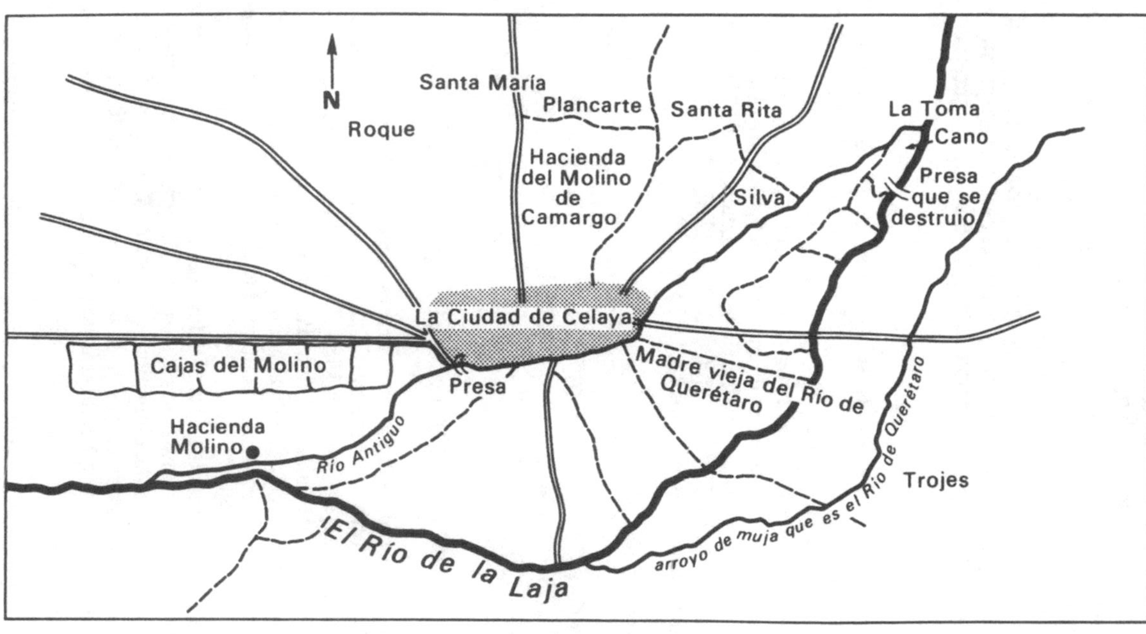

Figure 1.3 Celaya and environs depicted by Tresguerras. This interpretation of the original map shown opposite gives a panorama of the haciendas and ranchos around the city, and depicts the storage basins of the Hacienda del Molino. The flood course of the River Laja, often called the Rio de los Sabinos, is shown as "el rio antiguo". The River Queretaro is shown as an insignificant stream.

century, and a great many more were surely issued.[50] Around 1606, there was a burst of new grants; no fewer than fifty-five were issued in a single year.[51] Between 1607 and 1633, forty-one more vecindades were granted,[52] and four more were issued in the remainder of the colonial period, the last being in 1676.[53] The grants were never of more than two and one-half caballerías and were occasionally smaller in size. The restriction on the sale of the vecindad within ten years of the grant was strictly observed in the first decade of settlement, and some restriction on alienation appears to have remained in effect at least to the end of the sixteenth century.[54]

By good fortune, one can trace much of the evolution of land tenure generated by this profusion of small land grants. A remarkable document in the Franciscan archive in Celaya contains a summary of eighty-eight land sales before 1633; property descriptions in the document provide references to fifty-six more transactions.[55] Together with other sources,[56] the document provides a large, though fragmentary, sample of land transactions in the first sixty-two years of Celaya's existence. Not surprisingly, this sample reveals that the vast majority of land transactions in the period concerned small

---

[50]AGN-G, Vol. 2, Exp. 68; AFC-AP, P, leg. 1; AGN-M, Vol. 10, ff. 16v, 17, 56v, 69, 182, 183, 202v, 210v, 219; Vol. 11, f. 95v; Vol. 12, f. 156; Vol. 14, ff. 126v, 165v; Vol. 16, ff. 7, 23, 27v; Vol. 18, ff. 134, 144v, 216, 298, 298v; Vol. 19, ff. 105, 186. A document in the Franciscan Archive of Celaya contains numerous references to vecindades which are not recorded in the ramo de mercedes in the Archivo General de la Nación, probably because they were never confirmed by the audiencia. AFC-CC, E-2, leg. 2, and I, leg. 10. See also AGN-T, Vol. 674, Cuad. 2, ff. 1296, 1315 (Pedro Juárez and Francisco Couramuñez) and AGN-M, Vol. 18, ff. 143v, 288.

[51]AGN-M, Vol. 25, ff. 72-86, 90v, 91, 152.

[52]AGN-M, Vol. 26, ff. 66v, 67, 67v, 68v, 69, 109, 118, 118v, 174v; Vol. 28, ff. 263, 314; Vol. 30, ff. 30v, 105v; Vol. 31, ff. 204, 364v; Vol. 32, ff. 41v, 253v; Vol. 33, ff. 88, 244v; Vol. 35, ff. 123, 133, 136v; Vol. 36, ff. 4v, 5; Vol. 37, ff. 22v, 23, 68v, 177, 183, 184v; Vol. 38, ff. 8v, 13v, 92v; Vol. 39, ff. 68v, 136; AGN-T, Vol. 2760, Exp 1.

[53]AGN-M, Vol. 49, ff. 155v, 164 (1658-59); Vol. 50, f. 2v (1651); Vol. 58, f. 87 (1676).

[54]The Franciscan document described in the next paragraph mentions only nine transactions in the 1570s but records an active traffic in land beginning in 1580; there were six sales in 1580; six in 1581; and eight in 1582. Two attempted sales in the 1570s were declared invalid by the cabildo (sellers: Diego Pérez de Lemus and Gonzalo Díaz). Evidently, the restriction was lifted in 1580 by calculating the ten year period from the beginning of 1570. Later grants of vecindades continued to be subject to restrictions on sale. See AGN-M, Vol. 14, f. 178 (1588); Vol. 18, f. 288 (1594); Vol. 19, f. 186 (1594). But the fragmentary record of land grants and land sales does not permit us to determine the precise nature of the restriction or how it was enforced. The grants merely state generally that the grantee is subject to the obligations of other residents (e.g., "con las calidades se han concedido a los demas vecinos"). In the seventeenth century, this clause continues to appear frequently on grants, but there is no evidence as to whether or not it represented an effective restraint.

[55]AFC-CC, E-2, leg. 2. The following pages will be based on this document unless other additional references are given.

[56]For example, AGN-T, Vol. 674, Cuad. 2, ff. 1276-1360; AFC-CC, I, leg. 10.

parcels of land. There is a record of 148 land sales in which the size of the parcel is specified. Of this total, sixty-four involved the sale of a vecindad; fifty-seven involved the sale of smaller parcels--usually one or one and one- half caballerías in size; and twenty-three concerned the sale of larger properties.

This active market in small properties promoted a process of consolidation. The property descriptions and estate inventories in the Franciscan document record the consolidation of forty-one parcels into seventeen individual holdings. Many purchasers of small parcels mentioned in the document can be identified with previous purchases: nine purchasers accounted for forty-seven percent of the sales. The cumulative effect of such dealings in small properties over a twenty-five year period can be seen in the area along the southern bank of the Laja that was originally reserved as an ejido. In 1582 there were at least ten farmers in the area; in 1607, there were only three.[57]

The process of consolidation involved a large number of protagonists and favored the emergence of medium-sized haciendas. Some farmers appear to have purchased one or two caballerías to add to their vecindad.[58] A number of others acquired estates between five and ten caballerías in size; the eighty-six sales in the Franciscan document reveal eighteen acquisitions of this size. Our sample of sales transactions reveals only eight individual accumulations of over ten caballerías andonly one extensive estate, probably well in excess of twenty-eight caballerías.[59] In addition, five surviving estate inventories include an estate of twelve caballerías and another of twenty-three caballerías.[60] Vague property descriptions suggest the existence of a few other large holdings, but they were clearly exceptional. There were no extensive church properties in this early period, although the Franciscan college in Celaya possessed haciendas for a while in the seventeenth century.[61]

The process of consolidation was accompanied by the persistence of many small holdings, but the actual importance of these properties is obscure. About forty percent of the purchasers mentioned in the Franciscan inventory of transactions purchased property of two and one-half caballerías or less and cannot be identified with other holdings. The numerous grants of vecindades in the seventeenth century added to the number of property owners.[62] But our sample of transactions is limited and ends in 1633. Moreover, the early tithe records reflect a less numerous class of small farmers than these transactions and the record of land grants would suggest.[63] While the existence of many

---

[57]AGN-T, Vol. 674, Cuad. 1, f. 116v-120, Cuad. 2, f. 1215.

[58]For example, Juan Frayle and Cristóbal Venítez. See AGN-M, Vol. 10, ff. 41v-46.

[59]See AGN-T, Vol. 674, Cuad. 2, f. 1334v and AFC-CC, E-2, leg. 2 (Antonio Martínez de Contreras).

[60]See AFC-CC, I, leg. 10.

[61]In 1656 the Colegio de San Francisco appears as the owner of two haciendas, San Augustín with nine caballerías and San Nicolás with twelve caballerías. AGN-M, Vol. 49, f. 121.

[62]Only a few of the recipients of mercedes in the early seventeenth century appear in land transactions or in the tithe reports cited herein (e.g., Gonzalo Tello, Pedro Nuñez de la Roja, Alexo de Losa, Francisco de Aguilar, Pérez de Lemus), suggesting that many grants were given to landless residents who later sold or rented the land.

[63]El Obispado de Michoacán, p. 156; ACM, leg. 838.

small landowners in the first half of the seventeenth century cannot be doubted, one is left to speculate on their social importance.[64]

What role did water rights have in this evolving pattern of land tenure? In the 1570s, nine out of twelve grants of vecindades included a right to days of water, but after 1580, most grants do not mention water,[65] and after 1600, only two grants conferred water rights,[66] even though a number of these grants are described by proximity to the River Laja.[67] One begins to hear complaints of the inadequacy of two days of water as the normal allotment per vecindad,[68] and several farmers appealed to the audiencia in 1607 to halt usurpations of water by persons lacking formal grants.[69] Clearly, grants of days of water to new settlers served for a while to distribute more widely the existing supply, but the limit was soon reached beyond which the volume of existing water rights could not be further diminished. The relative importance of irrigated land declined as the settlement spread out from its nucleus along the River Laja and more irrigated land was cultivated. All five existing estate inventories, comprising property acquired in a series of transactions, include days of water, but only twenty-four out of eighty-eight transactions in the Franciscan document expressly mention the transfer of water rights.

There is unambiguous evidence of only one small sale of water rights separate from land in the sixteenth century.[70] In two other cases, vecinos sought administrative permission to transfer water rights.[71] The early settlers apparently were guided by the legal concept, prevalent in many parts of Spain, of water rights as interests adhering to particular parcels of land, and probably also by memories by Roble's early order forbidding the sale of water rights (which may itself have reflected this legal concept).

The cabildo made two grants of water rights independent of land; one of two days on the River Laja and another of four days on the River Querétaro. The latter went to Miguel Juan, a large landowner who had already received two days of water with his vecindad. In addition, two farmers applied successfully for water rights that other vecinos

---

[64]Compare: Brading (1978), p. 62; AGN-T, Vol. 192, Exp 1, f. 148 (division of ejido surrounding León in the seventeenth century).

[65]Between 1580 and 1600, five out of fifteen grants included water rights. See citations in footnote 50.

[66]AGN-M, Vol. 26, ff. 67, 67v.

[67]E.g., AGN-M, Vol. 31, ff. 204, 364v; Vol. 33, ff. 88, 244v.

[68]AGN-M, Vol. 14, f. 178.

[69]AGN-M, Vol. 29, f. 122.

[70]AFC-CC, E-2, leg. 2, item 14. A second sale may have occurred, but the text is unclear. The language describes five purchases of two days of water and the sale of ten days of water, but later passages reveal that the purchases were made in connection with purchases of land, suggesting that the sale of water rights may also have been made in a land sale. AGN-T, Vol. 674, Cuad. 2, f. 1291. For a later reference to the order forbidding sale of water rights, see AGN-T, Vol. 1175, Exp. 4, f. 67v.

[71]AGN-G, Vol. 2, Exp. 114 (Vecino asks permission to use two days of water on land other than that for which they were granted (1579); and AGN-M, Vol. 18, f. 134. (The grant of water rights to petitioner following another's renunciation of rights appears to have been a manoeuver to transfer water rights legally.)

had forfeited by selling or leaving their vecindad in violation of their obligation to occupy it for a ten year period.[72] Ponce de León was the most successful practitioner of this tactic. After complaining that three vecinos had sold their vecindades within one year after receiving them, he secured a grant of six days of water. His success had only small importance, however, since the rights were to water on the meagre flow of the River Querétaro.

In general, the control of water followed on the control of land. The examples of direct acquisition of water rights are few in number, and the amount of water affected was small. But the limited supply of water may have stimulated many of the sales and purchases that led to the ultimate shape of land tenure. In 1588, a settler observed that

> Ever since the villa of Celaya was founded, the vecinos have sold their vecindades and lands and water rights to each other because of the small quantity of water that was given to each one.[73]

One Luis Hernández amassed water rights along the River Querétaro by purchasing small parcels of irrigable land with related rights to days of water.[74] While the struggle for the control of water as such was of minor importance, the shortage of water may have contributed to the process of consolidation of rural estates in the early years of Celaya.

Wheat production continued at a relatively high level. In 1600 it was reported that over 30,000 fanegas were produced in the villa:

> They harvest in this town and its jurisdiction a quantity of more than 30,000 fanegas of wheat, which is milled into flour and carried and carted to the mines of San Luis, Guanajuato, Sichu, Tlalpuyahua and other places, for which purpose more than 1500 mules are employed in the town.[75]

But such an abundant harvest was exceptional in the seventeenth century. The tithe reports usually record production in the range of 12-24,000 fanegas, a level comparable to the 16-18,000 fanegas reported in 1580.[76] The haciendas and ranchos that account for this production varied from year to year, but we may safely infer that most of the wheat was grown on properties irrigated by the canal that was earlier described as the "upper" or "new" canal, originating in the dam later known as the dam of the Laboradores. Nearly all references to irrigation systems consist of the phrase "la acequia que viene al riego de

---

[72]AGN-M, Vol. 13, f. 114; Vol. 14, f. 178 (see also Vol. 18, ff. 268 and 288); AGN-G, Vol. 4, Exp 5.

[73]AGN-M, Vol. 14, f. 178.

[74]AGN-T, Vol. 674, Cuad. 2, ff. 1291-1317.

[75]Zavala (1947), p. 177.

[76]El Obispado en Michoacán, p. 156; ACM leg. 838, 860 (1661) and 861 (1663).

los labradores del Rio San Miguel" (Laja), or a close variation of these words.[77] At the head of the canal was an earthen dam. Although this dam was evidently effective enough to irrigate much of the valley, it could not be relied upon to provide water through the dry season; the water supply was known to fail as early as December.[78] The wheat fields, sown some time in October or November, were vulnerable to drought.

The needed reconstruction of the dam did not occur until an individual landowner took action. In 1634, Antonio de Abaunza, who owned a large unirrigated hacienda near the head of the canal,[79] saw an opportunity to secure a privileged position in the use of the water. In an application to the audiencia, he proposed to build a masonry dam in consideration for a grant of fourteen days of water and permission to build a flour mill at a favorable location near the head of the canal. The application was granted on the condition that the dam be built within one year. Abaunza evidently complied with this condition well enough to retain the right to operate the flour mill. In 1551, his successor, Diego de Soria Landin, secured permission to move the mill 500 varas upstream in the canal.[80]

The fact that a private landowner undertook this project suggests that the management of the irrigation system was shifting from municipal control into the hands of the principal water users. There are other similar indications. Abaunza's petition refers to "the shares established by the municipal government (ayuntamiento) in these allocations of water," but, instead of a water judge, it mentions "the man in charge of the dam," whose salary was apportioned among users. Water rights were now freely transferable so that landowners could negotiate among themselves modifications in the distribution of water. Two religious confraternities rented water rights. One Pedro de las Casas, who owned property near the periphery of the irrigation system, bought thirteen days of water in two transactions, but later sold most of his rights and died with only three days of water.[81]

The other major innovation in the seventeenth century was the construction of a masonry aqueduct to bring water into the town for domestic uses. The project was undertaken by the municipality but was carried to completion only with the intervention of a private party, this time the Convent of Carmen. The work began about 1637 with the object of connecting the fountain in the central plaza with an irrigation box two or three kilometers north of the city. Construction proceeded for more than ten years but was hindered by lack of funds. The Convent of Carmen kept the project alive with contributions that in 1647 totaled 14,000 pesos, as compared to 10,000 pesos then

---

[77]AFC-CC, E-2, leg. 7 (1655), and leg. 2, item 15; AFC-CSC, leg. "titulos" (1623); AGN-M, Vol. 26, ff. 118, 118v, 174v (1608). Two documents that speak of the "acequia principal" perhaps recognize implicitly the survival of a small canal on the River Querétaro. AGN-M, Vol. 28, f. 263; AGN-T, Vol. 187, Exp. 2, f. 100.

[78]In AGN-T, Vol. 187, Exp. 2, f. 100, a witness stated that the canal ran dry in December 1634. In an unrelated proceeding the same year, Antonio de Abaunza indicated with probably exaggeration that this was a normal occurrence. AGN-M, Vol. 39, f. 203v. See also AGN-T, Vol. 674, Cuad. 2, ff. 1118, 1214v.

[79]In 1631, he reportedly produced 4,000 fanegas of maize and fifty fanegas of beans. El Obispado de Michoacán, p. 156; AGN-M, Vol. 39, f. 203v.

[80]AGN-M, Vol. 49, f. 5v. See also AGN-T, Vol. 1175, Exp. 4, f. 60v.

[81]AFC-CSC, leg. "titulos." His property was later called the Hacienda San José near the Hacienda Tenería.

expended by the cabildo. Pressed for funds, the cabildo petitioned that its annual contribution for the cost of the drainage works in the valley of Mexico be reduced to permit further investment.[82] When the work was somehow completed, the cabildo shared with the Carmelites the expense of maintaining the system. In 1659, the Convent was given ten pajas of water in consideration for repairs costing 2,000 pesos, and in 1670 and 1672, it was given twelve caballerías of unoccupied land in exchange for further repairs amounting to 2,500 pesos.[83]

Celaya grew to be a provincial city of intermediate rank in the eighteenth century. Relying on ecclesiastical censuses, Morin calculates that the city and its Indian barrios had a population of 8,426 persons in 1772 and 9,217 persons in 1799. The population density of the region surrounding the city increased dramatically; two counts of the parish population, which included the Indian pueblos of San Juan de la Vega and Chamacuero to the north, show an increase in population from 5,381 in 1670 to 33,272 in 1760.[84] But the tithe data, though subject to administrative vagaries, suggest a much smaller increase in the level of production than one might expect from this increase in population. A sampling of fifteen tithe reports in the last half of the eighteenth century shows average wheat production of 23,400 fanegas, a modest increase over seventeenth century levels. Moreover, if we compare maize production for these years with a sample of ten years in the last half of the seventeenth century, we find an increase in production from 37,000 fanegas to only 61,000 fanegas.[85] The relatively stable level of wheat production evidently reflects the early exploitation of the hydraulic potential of the Laja, but the small increase in maize production is difficult to explain. It suggests, however, that there was relatively little intensified cultivation of land in the eighteenth century.

Celaya was a point of distribution for agricultural products, and a center of textile and leather manufacturing, known for its blankets, wool yarn, and soft leathers.[86] A small propertied class lived in elegant homes close to the central square,[87] while the indigenous population congregated in barrios at the fringe of the urban area and along a small arroyo to the south of the city called the River of the Sabinos (Figure 1.3). When H.G. Ward visited Celaya in 1827, he commented:

> Celaya, by the census of 1825, contains only 9,571 inhabitants; the streets are drawn as usual, at right angles, and the houses in the center of the town are well built; the suburbs are poor and miserable; but the great Plaza, one side of which is occupied by

---

[82]The distribution point (repartimiento) lay in the Hacienda Plancarte in the eighteenth century. AGN-T, Vol. 1362, Exp. 5, f. 6; AGN-A, Vol. 180, Exp. 6, f. 26. Velasco y Mendoza, relying on an anonymous history of the Convent of Carmen, gives the date of commencement as 1634; a document written in 1647 states that it was begun ten years earlier. Velasco y Mendoza, I, p. 114; AGN-T, Vol. 2682, Exp. 26.

[83]AGN-M, Vol. 52, f. 57; Vol. 56, ff. 66v-77.

[84]Morin, pp. 62, 73. Compare: Gerhard (1962), p. 22; Miranda, pp. 182-89.

[85]ACM, leg. 835, 838, 841, and 847.

[86]Moreno Toscano, passim.

[87]Consider the several elegant colonial residences that have survived to the present day in Celaya.

> the church of El Carmen, and the other by the convent of San Francisco, is really fine, and does credit to the taste of the architect (a native Mexican) by whom it was designed.[88]

Five religious orders, Carmelite, Augustinians, Mercedarians, Franciscans, and the hospital order of San Juan de Dios (*Juaninos*) maintained establishments in the city. Annexed to the Franciscan church was the Royal and Pontifical University of the Immaculate Conception of Celaya with faculties in grammar, humanities, rhetoric, arts, theology, and indigenous languages. Its most eminent graduate, the native Mexican to whom Ward refers, was a citizen of Celaya who resided there all his life: Francisco Eduardo Tresguerras, a distinguished architect who can stand comparison with the best contemporary architects of Europe.[89]

Although few municipal records have survived, we may surmise that, as in other small municipalities, members of the local landowning elite presided over the cabildo.[90] The glimpses into municipal affairs afforded by water cases suggests that important decisions were characteristically resolved by petitions to judicial authorities or to the central government. The cabildo disposed of a very modest assortment of revenues that amounted to only 2,076 pesos in 1775 from the following sources:

225 pesos--licensing of meat supplies
987 pesos--fees for storage of grain in the *alhondiga*
200 pesos--fees for use of central plaza
300 pesos--rental of water rights
403 pesos--rental of town lands

Revenues from these sources did not vary much from year to year but rose appreciably during the twenty year period from 1765 to 1784 for which we have records.[91]

---

[88]Ward, II, p. 421.

[89]Jiménez Jaime, pp. 126, 240. Humboldt remarked that the palace of Casa Rul in Guanajuato, designed by Tresguerras, "would adorn the finest streets of Paris and Naples." Quoted in Brading (1975), p. 395.

[90]For example, Manual Gómez de Linares, owner of Hacienda Roque (Velasco y Mendoza, I, p. 359 and ACM, leg. 841) and José Guadalupe Soria, Hacienda del Molinito (AGN-T, Vol. 2071, Exp. 1, ff. 1-7).

[91]1765-1308 pesos (AGN-A, Vol. 169, Exp. 20)
1766-1677 pesos (Ibid)
1767-1669 pesos (AGN-A, Vol. 218, Exp. 4)
1775-2076 pesos (AGN-A, Vol. 108, Exp. 13)
1779-1964 pesos (Ibid)
1781-2307 pesos (AGN-A, Vol. 169, Exp. 6)
1784-2197 pesos (AGN-A, Vol. 169, Exp. 2).

The countryside around Celaya was populated mostly by medium-sized haciendas between ten and twenty caballerías in size.[92] There were no great estates--only the largest haciendas had a population of over twenty-five men of Spanish descent--but some landowners held more than one hacienda; twenty individuals possessed forty-four haciendas in 1793.[93] Among the religious orders, the Carmelites administered the largest collection of properties.[94] Smaller ranchos were common, but they were marginal in importance and did not participate in the principal irrigation system. Tresguerras gives us a panorama of this landscape in a detail of one of his maps (Figure 1.3). We can see a patchwork of haciendas north of the city within the area of the irrigation system. To the west, a large estate, the Hacienda Roque, dominates the scene, and, along the river Laja, there is a string of small ranchos.[95]

---

[92]For example, San Lorenzo Aguas Calientes, 5 caballerías in 1720 (AFC-CSC, leg. "pleitos"); Lo de Muñiz and Tenería were considered as 1 hacienda of 20 caballerías in 1758 and as 2 haciendas of 20 caballerías c. 1737 (ACM, leg. 838, Exp. 578, and AFC-CC, L-4, la pte, ff. 15, 22v); San Nicolás de Plancarte, 16 caballerías in 1742 and 11 caballerías in 1789 (ACM, leg 825, exp 368, and leg 852, exp 1256); Guadalupe, 19 caballerías in 1763 (ACM, leg. 834, exp 462); San Miguelito, 12 caballerías in 1707 (AGN-T, Vol. 237, Exp. 6, f. 5); Santa María, 12 caballerías, Rancho de Arroyo Hondo, 2-1/2 caballerías, Rancho de la Resurrección, 3 caballerías in 1727 (AGN-T, Vol. 618, Exp. 1, ff. 55-65); San Isidro de Sarabia, 20 caballerías and unnamed rancho, 2-1/2 caballerías in 1714 (AGN-T, Vol. 238, Exp. 1, f. 101); Rancho lo de Pérez o La Laja, 3-1/2 caballerías in 1741 (AGN-T, Vol. 621, Exp. 1, f. 106); San José de la Gavia, 25 caballerías, San Elias, 32 caballerías in 1716 (AGN-M, Vol. 70, f. 29v); Santa Rosa, 7 caballerías in 1699 (MNA.AM-Qro, roll 27, Registro de la Propiedad, no. 24).

[93]AGN-P, Vol. 26, f. 273 et seq. See Haciendas San Nicolás, Morales, Concepción, and Molino de Soria. Also: Morin, p. 212.

[94]For example, Augustinians: San Nicolás and Morales in 1793 (AGN-P, Vol. 26, f. 273); Jesuits: Camargo and del Molino (Velasco y Mendoza, I, p. 215); Carmelites: San Elias in 1803 (AGN-T, Vol. 1345, Exp 3, f. 1), Santa Rosalia and Morales in 1780s (ACM, leg. 841), Jofre, la Gavia, Celecio, Lo de Estrada in 1793 (AGN-P, Vol. 26, f. 273). In addition, a pious fund sponsored by the Carmelites, El Santuario de la Santísima Cruz (see AGN-T, Vol. 1168, Exp. 3, f. 6), owned Santa Teresa and Tenería in 1793 (AGN-P, Vol. 26, f. 273).

[95]Many but not all of the ranchos along the river were rented from larger haciendas. See Rancho lo de Pérez and Resurrección in footnote 92 above and AGN-T, Vol. 1390, Exp. 3, f. 20 (se notificaron a los dueños, arrendatarios y demas posedores de los haciendas contiguos al río).

The largest wheat haciendas--they were only about seven in number--all lay within the irrigation system of the dam of the Labradores.[96] The Haciendas Mendoza, Plancarte, Santa Rosa, Santa María, Camargo, Santa Rita, and Roque formed the bulk of the system and occupied most of the valley to the north and northwest of Celaya. A tributary canal led west to the very important Hacienda Roque, about thirteen kilometers from the dam.[97] Not far from the head of the canal, the Haciendas San Miguelito and Tenería periodically rented water.[98] Four other haciendas (and very probably more) possessed minor shares of water rights.[99]

A significant portion of the water rights was held by organizations that did not own irrigated land.[100] Two confraternities associated with the church of San Francisco together held eight days of water.[101] The pious fund (*obra pía*) of San Andrés had five days of water,[102] and the Franciscan University had two days of water.[103] The city enjoyed a right to a continuous supply from the irrigation box on the Hacienda Plancarte

---

[96]Production statistics: ACM, leg. 841. Records of water rights: Santa María, 19 days of water in 1727 (AGN-T, Vol. 618, Exp. 1, f. 55v); Camargo, 18 days of water c. 1707 and 10 days of water in 1749 (AGN-T, Vol. 237, Exp. 6, f. 25v, and AFC-CC, E-4, leg. 95); Santa Rita, 12 days of water in 1682 (AFC-CSC, leg. "titulos"); San Nicolás de Plancarte, 22 days of water in 1716, 19 days in 1742 and 10-1/2 days in 1789 (ACM, leg. 825, Exp. 368, and leg. 852, Exp. 1256); Santa Rose, 12 days of water in 1699 (MNA.AM-Qro, roll 27, Registro de la Propiedad, no. 24); Mendoza, location of irrigation box in 1774 and rental of water in 1746 (AGN-A, Vol. 108, Exp. 2, f. 1, and AFC-CC, C-3, leg. 1, f. 49). In addition, it is possible to identify the owners of the Haciendas Camargo, Santa Rita, Santa Rosa and Mendoza as being present at the meeting of irrgators in 1788. Compare: AGN-T, Vol. 1168, Exp. 3, f. 6, and ACM, leg. 841.

[97]AFC-CC, T-2, leg. 4 (1706); AGN-A, Vol. 218, Exp. 4 (The city rented water to owner of hacienda after 1768); AGN-T, Vol. 1168, Exp. 3, f. 6 (Owner present at meeting of irrigators in 1788).

[98]San Miguelito rented eight days of water c. 1718. AGN-T, Vol. 353, Exp. 2, f. 10; AFC-CC, C-3, leg. 1. I have a record of only one year in which the Hacienda Tenería rented water (AGN-T, Vol. 353, Exp. 2, f. 13v), but its proximity to the canal and the fact that it was described as being "de temporal" (AFC-CC, L-4, la pte, f. 83) and produced a large wheat crop in some years but not in others suggests that it periodically rented water. ACM, leg. 841.

[99]San José de la Gavia, 4 days of water in 1716 (AGN-M, Vol. 70, f. 29v); San Gerónimo, 7 days of water in 1739 (AGN-T, Vol. 353, Exp. 2, f. 5, and AGN-T, Vol. 450, Exp. 5, f. 13v); La Concepción (AFC-CC, I, leg. 12, f. 83); San Antonio Mujica, 1 day of water in 1714 and owner present at meeting of irrigators in 1788 (AGN-T, Vol. 238, Exp. 1, ff. 2v, 101, and AGN-T, Vol. 1168, Exp. 3, f. 6).

[100]See AFC-CC, L-4, la pte, Cuad. 2, f. 6 (one day of water pledged as a security for purchase price of land).

[101]AGN-T, Vol. 450, Exp. 5, f. 17v; AFC-CC, C-3, leg. 1, *El Obispado de Michoacán*, p. 156.

[102]AGN-T, Vol. 353, Exp. 2, f. 5.

[103]AFC-CC, I, leg. 12 (1700).

and rented out six additional days of water from the irrigation box on the Hacienda Mendoza near San Juan de la Vega.[104] This municipal share was apparently determined in seventeenth century litigation with the farmers of Celaya, but the documentary records provide no details.[105] In the course of the century, the market rental of a day of water rose from twenty-five pesos a year to fifty pesos a year.[106]

The market in water rights permitted enterprising farmers to expand their irrigated fields. About 1696 Nicolás de Molina bought twelve unirrigated caballerías along the river Laja, rebuilt his barns, and placed additional land under cultivation. He succeeded in renting six days of water from the city despite objections that his property was too heavily mortgaged and rented two additional days of water from a neighboring hacienda, San Geronimo, which itself rented the water from the obra pia of San Andrés. In 1718, he outbid his neighbors and rented the water directly from the obra pia. In retaliation, the owner of the Hacienda San Gerónimo denied him the passage of the water across his property. Molina subleased his water rights to another hacienda and sought judicial relief. Within a year he secured an order upholding his right to conduct the water over the Hacienda San Geronimo to his property.[107]

In the eighteenth century, there were two major irrigation boxes in the canal system of the dam of the Labradores: one on the Hacienda Plancarte and the other on the Hacienda Mendoza.[108] Both had to be properly maintained and administered. During the rainy season, the needs of flour mills had to be reconciled with those of the farmers.[109] The number of days of water granted periodically in the sixteenth century to the farmers along the original canal surely bore no relationship to the total number of days of irrigation needed for the wheat crop or even any convenient fraction thereof. But somehow a method of rotation was devised that preserved the system of individual rights to days of water and irrigated the farmers' fields when they needed it. A system of rights to continuous supplies of water was engrafted on the system of rotation. The Hacienda

---

[104]AGN-A, Vol. 108, Exp. 13, f. 1 (refers to "el arrendamiento de los seis días de agua sobrantes de los nueve días que tiene esta ciudad."); AGN-T, Vol. 1169, Exp. 1, f. 78 (refers to ownership of ten days). See also AGN-A, Vol. 108, Exp. 2, f. 1; Vol. 180, Exp. 6, f. 25.

[105]AGN-T, Vol. 237, Exp. 6, f. 6.

[106]1700--25 pesos (AFC-CC, I, leg. 12)
1707--25 pesos (AGN-T, Vol. 237, Exp. 6, f. 6v)
1718--40 pesos (AGN-T, Vol. 353, Exp. 2, f. 5)
1727--30 pesos (AGC-CC, C-3, leg. 1)
1739--35 pesos (AGN-T, Vol. 450, Exp. 5, f. 12)
1746--50 pesos (AFC-CC, C-3, leg. 1)
1775-9--50 pesos (AGN-A, Vol. 108, Exp. 13)

See also AGN-T, Vol. 1175, Exp. 4, f. 62 (c. 1772): "cada dia de agua se evalua en 1500 o 2000 pesos segun el marco que les cabe."

[107]AGN-T, Vol. 237, Exp. 6; Vol. 353, Exp. 2.

[108]AGN-A, Vol. 108, Exp. 2, f. 1; Vol. 180, Exp. 6, f. 25.

[109]AGN-M, Vol. 72, f. 174v.

San José de Gavia, for example, possessed rights both to four days of water and to a continuous supply, one naranja of water.[110]

The sparse available information indicates that the elaborate adjustments needed to administer the system were worked out at an early date, probably in the seventeenth century, and maintained largely by custom and self-enforcement by individual haciendas. New problems were handled on an ad hoc basis by meetings of the principal landowners. In eighteenth century documents, one finds many vague references to customs and judicial decrees but no mention of municipal ordinances, systems of assessment, or compacts among farmers.[111] Documents of the sort that ought to reveal something of the workings of the system do not contain the barest allusion to a water judge or other enforcement official. But we know that keys were used to open the gates of the irrigation boxes; they were undoubtedly entrusted to specially appointed persons.[112] It is perhaps safest to infer that irrigation officials existed but had a minor role. Certainly, owners of water rights took action individually to protect their interests. The cabildo employed an Indian to guard the irrigation box at the Hacienda Plancarte, and the Hacienda Santa María employed watchmen to guard its canals.[113] Individual haciendas were understood to have an obligation to maintain the length of canals within their properties.[114]

The only record of a meeting of landowners concerning irrigation tends to corroborate the picture of a system lacking a well-organized institutional structure and managed informally by a small group of the largest property owners. In 1788 seven representatives of the largest haciendas met to consider repairs of the dam of the Labradores. They agreed that each party would contribute labor for the repair of the dam and that all costs would be verified so that they could be prorated among the haciendas.[115]

The wheat haciendas continued to be vulnerable to drought toward the end of the dry season.[116] Despite the inevitable loss of water in conduction, several of the largest haciendas were at the southern limit of the irrigation system. Water shortages led to

---

[110]AGN-M, Vol. 70, f. 29v.

[111]E.g., AGN-T, Vol. 1169, Exp. 1, f. 30v ("de manera que las varias diferencias que se ofrecieron entre los mismos labradores, se dieron en los tribunales muchas providencias"); AGN-T, Vol. 237, Exp. 6, f. 7v ("al tiempo y cuando es uso y costumbre entre los labradores").

[112]AGN-A, Vol. 169, Exp. 20.

[113]AFC-CC, L-4, la pte, Cuad. 4; AGN-A, Vol. 108, Exp. 13, f. 2.

[114]AFC-CC, L-4, la pte, Cuad. 4; AGN-T, Vol. 450, Exp. 5, f. 22.

[115]The text mentions only representatives of the most important haciendas: Roque (Tomás Fernández Cavada), Santa María and Santa Rosa (Miguel Galván), Mendoza (Juan Cavallero), San Antonio (Pedro Muxica), Plancarte (Joaquín Márquez) and Santa Rita (Luis Muxica) and the Carmelites. AGN-T, Vol. 1168, Exp. 3, f. 6.

[116]José Rodríguez, an elderly ejidatario in the area, told me that, before the construction of the dam of La Begoña, the dam of the Labradores, which was essentially the same as the colonial dam, provided the largest farms with a flow at only 70 liters/sec during an irrigation turn. See also AGN-T, Vol. 1168, Exp. 3, f. 33v.

litigation with upstream users near San Miguel El Grande.[117] But there was still some expansion of irrigation through the storage and later use of the summer flow of the river. This activity, which was never associated with formal grants of water rights, is well described in a document that notes the inadequacy of the flow in the dry season and explains:

> It is well known that for this reason many haciendas and landowners have undertaken to build and put into use through much labor and heavy investments new irrigation systems along the river to channel the water during the months of June, July and August when it rains most abundantly to holding ponds, dams and reservoirs that they have built on their haciendas. The water is stored in these holding ponds and dams to be used in the cultivation and irrigation of wheat, chile and other crops in the dry season.[118]

The storage of flood water for irrigation may, of course, have been an early practice,[119] but it is only in the eighteenth century that one finds evidence of the elaborately sculptured landscape--visible in the early twentieth century--of dikes and storage basins designed to facilitate this process.[120]

At the end of the eighteenth century, these storage basins were found throughout the central irrigation system of Celaya.[121] The quantity of water stored was such that leakage from the basins was a source of concern to townspeople. The principal wheat haciendas received water channeled from the dam of the Labradores, but there were many efforts to conduct river water to storage basins by other means. Around 1780, the owner of the Hacienda Santa Rita constructed a dam on the Laja (soon destroyed by high water) to divert water to the eastern portion of his hacienda. To the south, a second small dam was built around 1790 to supply the rancho of Cristóbal Cano. Elsewhere there were simple ditches excavated in the river bank, not associated with dams, intended to capture high waters. One finds mention of these ephemeral irrigation intakes (sacas) in the lands of the "Martínez" near San Juan de la Vega and in the ranchos of Aguirre, Hurtado, and the Laja on the west bank near the city.[122] The most carefully engineered of these irrigation systems--and the most offensive to other residents--was the canal of the

---

[117]There was a long history of irrigation on the River Laja above Celaya. AGN-M, Vol. 16, f. 117; Vol. 58, f. 82; Vol. 70, f. 93v; AGN-T, Vol. 1169, Exp. 1, Cuad. 5, f. 61. Two new dams were built around 1750 on the Hacienda Cieneguilla. AGN-T, Vol. 1169, Exp. 1, Cuad. 5, ff. 16, 126; Vol. 1175, Exp. 4, f. 73; AGN-M, Vol. 81, ff. 3v, 55v.

[118]AGN-T, Vol. 1175, Exp. 4, f. 62v.

[119]AGN-T, Vol. 2705, Exp. 3, f. 1; (Reference to jaguey in area generally devoted to irrigation, 1632); AGN-M, Vol. 10, f. 3 (Apparent reference to flood plain farming, 1573).

[120]E.g., AGN-T, Vol. 2767, Exp. 3, Cuad. 5, f. 4v.

[121]ACM, leg. 852, Exp. 1256, f. 48 (Plancarte, 1789); AGN-T, Vol. 2072, Exp. 1, Cuad. 8, f. 99 (Reference on map to cajas of Camargo, Santa Rita, Plancarte, etc., 1810).

[122]AGN-T, Vol. 621, Exp. 8, f. 105 (1741); Vol. 1333, Exp. 4, ff. 15v-20; Vol. 1390, Exp. 3, f. 21v; Vol. 2071, Exp. 1, ff. 4, 44 (Luis Muxica).

Hacienda del Molino built in 1789, which conducted water southward toward the city itself.[123] There is curiously little record of irrigation on the east bank, but the Hacienda Trojes on the far side of the river south of the city built a system of storage basins near the end of the century.[124] The production in 1763 of a small quantity of trigo aventurero, a term applied to wheat sown in the autumn and not thereafter irrigated, on the Hacienda Guadalupe also suggests some rudimentary use of flood waters.[125] Away from the Laja itself, at least two haciendas built dams on small arroyos.[126]

The settlement of Celaya had long been traversed by canals. Early in the eighteenth century, water from the dam of the Labradores supplied a flour mill at the edge of the city. The flow from springs in a swamp to the north of the city, augmented by leakage from the municipal aqueduct,[127] fed another canal that received various municipal wastes and the overflow from the central fountain[128] (see Figures 1.3 and 6.7). In its passage through the city, the canal irrigated the gardens of the Carmelite, Augustinian, and Franciscan churches.[129] It served the function of that important feature of many colonial towns, the municipal canal, but it became an object of contention among private parties. The Carmelites secured a grant of rights to the spring water in the swamp and twice attempted to channel the water to the Hacienda Estrada to the west.[130] The Franciscans wanted to retain the original course of the canal to irrigate their garden.[131] The cabildo wished to reduce the level of water in the swamp for health reasons (the neighborhoods nearest the swamp were known to have the highest incidence of "chills") and vainly asserted a municipal right of domain over the canal.[132] Finally, the Hacienda del Molino wanted the water for irrigation. In the interminable legal proceedings

---

[123]AGN-T, Vol. 2767, Exp. 3, e.g., Cuad. 2, ff. 1v, 27, Cuad. 4, f. 3, Cuad. 5, f. 3v.

[124]Morin, p. 289.

[125]ACM, leg. 834, f. 14.

[126]There were small masonry dams on the Hacienda Escojido (AGN-T, Vol. 548, Exp. 2, f. 197v) and Morales (AGN-T, Vol. 1020, Exp. 1, Cuad. 2, f. 37v). The latter was used to produce a small quantity of wheat (ACM, leg. 841). The few surviving inventories of haciendas contain further allusions to storage basins, norias and ditches leading from arroyos. AGN-C, Vol. 73, Exp. 3, ff. 16-22v; AGN-T, Vol. 514, Exp. 1, Cuad. 2, f. 19v (map); Vol. 1020, Exp. 1, f. 26v.

[127]AGN-T, Vol. 1362, Cuad. 1, f. 14v, Cuad. 5, f. 14.

[128]AGN-T, Vol. 2071, Exp. 1, f. 128; Vol. 2072, Exp. 1, Cuad. 8, f. 94 et seq.

[129]Zamarroni, II, p. 135; AGN-T, Vol. 1362, Cuad. 1, ff. 17v-35v, Cuad. 3, f. 6v.

[130]AGN-M, Vol. 72, f. 269; AGN-T, Vol. 1362, Cuad. 1, ff. 2-35v, Cuad. 3, f. 4v, Cuad. 4, f. 24; Vol. 2074, Exp. 2, f. 54.

[131]AFC-CC, D, leg. 19; AGN-T, Vol. 1362, Cuad. 1, ff. 18v, 23, Cuad. 3, f. 92v, Cuad. 12, f. 101; Vol. 2074, Exp. 2,
f. 50v.

[132]AGN-T, Vol. 1362, Cuad. 3, ff. 5, 92v, Cuad. 12, f. 48 (El barrio de Cienega era el que sufrio por lo regular la enfermedad de frios por las aguas estancadas.");
Vol. 2074, Exp. 2, f. 90v.

generated by this tug-of-war, it was curiously the Hacienda del Molino that consistently prevailed.[133]

The municipal aqueduct delivered a small and frequently interrupted supply of water to a large fountain in the central plaza and two smaller fountains at nearby convents.[134] It was plagued by problems of maintenance and was closed during the summer when it was susceptible to siltation, and the municipal canal provided an alternative water supply. In 1707, the disrepair of the aqueduct was said to cause "shortages and great hardship";[135] for periods between 1775 and 1782, the supply failed completely;[136] and in 1800, an expert calculated that 32/33ds of the water was lost in conduction. When the fountains ran dry, the townspeople were forced to resort to unsanitary well water.[137]

The cabildo responded characteristically by trying to enlist the efforts of private landowners in the maintenance of the aqueduct. The city's six days of water were rented on the condition that the recipient keep the aqueduct in proper repair. The obligation seems to have been taken seriously in the early years of the century; members of the cabildo criticized the performance of José Camargo but questioned the financial ability of Nicolás de Molina to do better.[138] Later, the cabildo devoted a major portion of the municipal budget to an unavailing effort to maintain the system. During the seventeen years for which there is record of municipal expenditures, the cabildo nearly always spent funds to clear and repair the aqueduct; in five years it spent more than 400 pesos and in nine years, more than 200 pesos.[139]

Like some other towns of the Bajío, Celaya was plagued by floods. During heavy rains, the Laja would break from its normal channel and follow a flood course leading to the city (Figure 1.2). The Indian barrios on either side of this water course were the first to be affected by any high water and the most severely devastated by major floods.[140] Periodically, the remainder of the city, including the central plaza and the Convent of San Francisco, also suffered inundations. Around 1800, a citizen lamented "the floods that Celaya suffers year after year which destroy workshops and churches and make the streets

---

[133]AGN-T, Vol. 1362, Cuad. 1, ff. 1-23v, Cuad. 3, ff. 5, 86v-92v, Cuad. 6, f. 19, Cuad. 12, f. 101.

[134]AGN-M, Vol. 52, f. 57; AGN-T, Vol. 237, Exp. 6, f. 4; Vol. 1168, Exp. 3, f. 32v.

[135]AGN-T, Vol. 237, Exp. 6, ff. 6v-7v; AGN-A, Vol. 108, Exp. 13.

[136]AGN-A, Vol. 108, Exp. 13; Vol. 180, Exp. 6, f. 28.

[137]AGN-T, Vol. 1362, Cuad. 6, f. 6.

[138]AGN-T, Vol. 237, Exp. 6, ff. 1-10v.

[139]AGN-A, Vol. 108, Exp. 1, 2, and 13; Vol. 169, Exp. 1, 2, 6, and 20; Vol. 180, Exp. 3 and 6; Vol. 218, Exp. 1 and 4.

[140]AGN-T, Vol. 2170, Exp. 1, Cuad. 1, f. 30, Cuad. 3, f. 16v; AGN-A, Vol. 180, Exp. 6, f. 24v.

impassible."[141] The chance survival of historical documents has left a record of damaging floods in 1692, 1750, 1753, 1767, 1774, 1781, and 1802.[142]

The disastrous flood of 1692 prompted the cabildo to secure authorization from the viceroy to build a dike north of the city. The viceroyal order directed the Spanish vecinos to make a specified monetary contribution and the Indian residents to provide labor. According to the late archivist of Celaya, who kept these documents in his personal possession, the work was actually carried to completion.[143] But except for a few minor expenditures,[144] the cabildo thereafter neglected the problem of flood control until it came into conflict with the owner of the Hacienda del Molino near the end of the next century.

The canal of the Hacienda del Molino traced part of the route by which flood waters invaded the city and was thought to aggravate the traditional flood hazard. Roused by this perceived threat, the cabildo order in 1790 that landowners remove dams and irrigation intakes on the Laja:

> The owners, renters and other persons in possession of land along the bank of the river... within eight days...shall close firmly with stakes and pressed earth the headgates and ditches on their property along the river, dike areas subject to flooding and also remove masonry dams located in the river.[145]

Despite the broad language of the order, the cabildo did not intend to order removal of the dam of the Labradores; only the farmers on the Laja close to the city were formally notified of the order.[146] But the cabildo, composed of an elite that commonly resided in the city, had moved a full circle from the early days of Celaya as a settlement of irrigation farmers. As an ally of the cabildo, Francisco Eduardo Tresguerras argued that the argicultural benefits of irrigation were of less importance than improved living conditions in the city.[147]

The cabildo applied to the central government for authorization to remove the offending dams and ditches. On January 20, 1791, the audiencia ordered that the alcalde mayor execute forthwith the resolution of the cabildo "to remove the dams and close the

---

[141]AGN-T, Vol. 1390, Exp. 3, f. 7v.

[142]AGN-A, Vol. 108, Exp. 2; Vol. 180, Exp. 6; AGN-T, Vol. 1362, Cuad. 12, f. 38; Vol. 1390, Exp. 3, f. 34v; Zamarroni, I, p. 151.

[143]Zamarroni, I, p. 151. Zamarroni believed that the cabildo considered building a major flood prevention dam near San Miguel Allende. His interpretation rests on the phrase in the petition of the alcalde mayor, "presa rio arriba." Although the text is obscure, I think it unlikely that anything other than a dike was envisaged.

[144]AGN-A, Vol. 108, Exp. 2; Vol. 180, Exp. 6, f. 24.

[145]AGN-T, Vol. 2071, Exp. 1, Cuad. 1, f. 1.

[146]AGN-T, Vol. 2171, Exp. 1, Cuad. 1, ff. 3v-8v; Vol. 1390, Exp. 3, f. 22v.

[147]AGN-T, Vol. 2071, Exp. 1, Cuad. 1, f. 30.

headgates on the river with the costs being to the account of the farmers themselves."[148] But the order was suspended on March 23 of the same year, and the audiencia directed further investigation. For fourteen years, the matter was alternatively abandoned and revived,[149] but in 1805, the cabildo was finally disposed to take action. Relying on the suspended order of 1791, it directed Tresguerras to destroy the irrigation system of the Molino by filling in the headgates and demolishing a small diversion dam. The cabildo later ordered the destruction of a second dam serving a rancho of Cristobal Cano near the headgates of the Hacienda del Molino.[150]

The cabildo clearly acted without proper authority and would pay the price for its independence.[151] The owner of the Hacienda del Molino quickly secured a provisional order directing the cabildo to open the canal, and the following year he obtained a viceroyal decree that made the members of the cabildo personally liable for the damages he had suffered. The cabildo attempted to carry on the fight, arguing the Hacienda del Molino's lack of title to the water. Tresguerras was moved to produce the masterful drawing reproduced in part as Figure 1.2. There is no record of whether the cabildo actually compensated the owner of the Hacienda del Molino for his loss, but the hacienda did continue to use the canal for irrigation.[152]

The leaders of Indian barrios were enlisted by both sides of this controversy. The barrios were exposed to flooding, but, with the permission of the Hacienda del Molino, they used the water from the canal for irrigation of gardens and corn fields that lined the River of the Sabinos.[153] The Indian population near Celaya consisted of Otomí and other groups who migrated into the region during the period of early Spanish

[148]AGN-T, Vol. 1390, Exp. 3, f. 8v. See also ff. 22v, 26.

[149]AGN-T, Vol. 1390, Exp. 3, ff. 27-36v; Vol. 2071, Cuad. 2, f. 5v; Vol. 2072, Exp. 1, Cuad. 1, f. 9, Cuad. 5, f. 13, Cuad. 8, f. 21.

[150]AGN-T, Vol. 1390, Exp. 3, f. 39-42; Vol. 2071, Exp. 1, Cuad. 1, ff. 28-41, 47-59; Vol. 2072, Exp. 1, Cuad 5, f. 28.

[151]Following a serious flood in 1750, the cabildo of Irapuato acted with prudence that the officials of Celaya would have been well advised to have imitated. The cabildo appointed experts to prepare a report that was sent to the fiscal de la civil of the audiencia. The fiscal reserved for further study a plan for diverting the River Silao to a new channel and ordered that a dike be built at a particularly vulnerable point on the river bank. Construction of the dike began immediately; costs were assessed against property owners, and Indians in the city's barrios were ordered to work at one-half the normal wage for field labor. Unfortunately, the dike did not protect the city from another flood in 1756. After commissioning another report by two experts, the cabildo ordered that a dam be built to divert the river away from the city and that the costs be paid by the haciendas that would benefit from the diversion. Wealthy landowners agreed to do the work; the dam effectively alleviated the flood hazard, but, because it channeled part of the river's flow to a new course, it eventually led to conflict between the haciendas near the city and the haciendas along the new course of the river. AGN-T, Vol. 1166, Exp. 1, ff. 1-77.

[152]AGN-T, Vol. 2071, Exp. 1, Cuad. 1, f. 68; Vol. 2072, Cuad. 8, ff. 62-94; Vol. 2073, Exp. 1, ff. 3-50; Velasco y Mendoza, I, p. 18.

[153]AGN-T, Vol. 1390, Exp. 3, ff. 28-31; Vol. 2071, Exp. 1, Cuad. 2, f. 31v; Vol. 2072, Exp. 1, Cuad. 5, f. 29.

colonization.[154] Unlike the Spanish settlers, they were not given water rights. In 1576 the cabildo of Celaya expressly forbid Indians from using the irrigation systems of the town,[155] and in 1594, the audiencia denied the Indians of San Gerónimo, a village near the San Juan de la Vega, an application for water rights.[156] San Juan de la Vega itself seems never to have participated in the irrigation system of the dam of the Labradores.[157] Only in 1710 is there a record of an Indian barrio receiving rights to water from the swamp.[158]

In Apaseo, Indians could also be found near the margin of irrigation systems at the sufferance of the owners. On the hacienda of the mayorazgo, a tributary ditch led to a collection of gardens and orchards where a group of laborers on the hacienda had settled. When their number became too great, the owner succeeded in expelling some, but not all of them.[159] But on the northern side of the river, the Indian pueblo retained a portion of the water rights originally ceded in the compact with Bocanegra.[160] In sixteenth century litigation, the pueblo had secured a decree recognizing its right to one-half of the volume of the river, but it conceded part of this share to a hacienda, known as the Hacienda de la Comunidad, that rented a large section of pueblo land.[161] At the beginning of the eighteenth century, the Indian community of Apaseo is reported to have used one-fourth of its half share of the river for its gardens and for domestic purposes; the remainder was used to irrigate small fields of wheat or was given to the Hacienda de la Comunidad. The Indians did not have assigned days for irrigating their land, but rather "when needing irrigation for their garden or small field, they would take the water, and when they finished, they would pass it to the vecino or Indians who then needed it."[162]

The renter of the Hacienda de la Comunidad in 1720 claimed only the right to water remaining after the Indians had used what they needed (el remaniente),[163] but a later renter of the hacienda fell into conflict with the Indian community. He was accused of altering the ditches at the expense of the Indians so as to irrigate not only the Hacienda

[154]AGN, Indios, Vol. 3, Exp. 936, f. 227; AGN-T, Vol. 487, Exp. 4; AFC-AP, P-1, estante 3, leg. 26; AFC-CC, E-2, leg. 7.

[155]AGN-T, Vol. 674, Cuad. 2, f. 1156v.

[156]AGN-T, Vol. 1169, Exp. 1, Cuad. 4, ff. 7v-31v; Vol. 1175, Exp. 4, f. 57.

[157]See ACM, leg. 841.

[158]Zamarroni, II, f. 135.

[159]AGN-T, Vol. 972, Exp. 8, ff. 1-2.

[160]AGN-T, Vol. 674, Cuad. 2, ff. 1115, 1198-1206v, 1251-64. The Indians were vindicated by an order dated February 8, 1608 that reinstated the previous order of the alcalde mayor in their favor. See also AGN-T, Vol. 187, Exp. 2, f. 19.

[161]AGN-T, Vol. 383, Exp. 4, f. 1; Vol. 674, Cuad. 1, f. 49, Cuad. 2, ff. 1206v-1264.

[162]AGN-T, Vol. 187, Exp. 2, f. 140. See also ff. 150v, 217v.

[163]AGN-T, Vol. 383, Exp. 4, ff. 1-136. For further reference to mill, see AGN-M, Vol. 38, f. 8v.

of the Comunidad but additional lands.[164] The Indians also suffered a minor theft of water at the hands of one Juan García de Alarcón. In 1628,the Indians had sold a garden plot to one Pedro Ruiz with rights to two days (dos días) of water. García de Alarcón, the heir of Ruiz, claimed twelve days of water by the simple means of writing an "e" after the "s" in the original document ("dose", a common archaic spelling of twelve). Although the forgery was plainly exposed, the audiencia gave him the right to four days of water that he channeled to newly acquired land downstream.[165]

At the end of the colonial period, the physical form of the irrigation systems near Celaya, despite some new investment in storage basins and diversion dams, retained much of the early pattern formed in the sixteenth century, and the Indian pueblo of Apaseo clung to some of its original water rights. But the practice of irrigation had declined in importance in an increasingly complex society. The irrigation systems had become largely the possession of a small collection of wheat-growing haciendas. The areas of tillage extended well beyond the reach of irrigation, and urban residents competed with irrigated haciendas over the use of water.

[164]AGN-T, Vol. 2901, Exp. 36; Vol. 2947, Exp. 100.

[165]AGN-T, Vol. 187, Exp. 2, ff. 1-57, 129-263; Vol. 383, Exp. 4, f. 158.

2
Salvatierra

After the conquest, the Spanish presence spread quickly to the valley of Acámbaro, an outlier of the Bajío along the River Lerma, that lay just within the region of sedentary Indian culture of the Tarascan Empire. If we can trust an early account, three Franciscan friars arrived at the site of San Francisco de Acámbaro in 1526 following a successful military expedition fought by Otomí allies of the Spanish invaders. The friars set about to give the indigenous settlement a Spanish form by laying out a grid of streets, building a chapel and later a hospital, and assigning houses for the caciques. Water supply was an immediate concern. Although the town lay near the bank of the Lerma, the friars looked toward an arroyo in the nearby Cerro de Ucareo for a manageable supply of water that could be brought to the town. We are told that one of the friars worked "all the year of 1527" building a canal from the arroyo to the town.[1]

About fifty kilometers to the northwest, Yuririapúndaro was an important Tarascan frontier town that was early adjudicated to the crown. Basalenque puts its pre-Hispanic population at 6,000. The town was assigned to the Augustinians, and the energetic Fray Diego de Chávez of that order evidently arrived around 1550.[2] According to Matias de Escobar, he soon reordered the town in conformance to a Spanish pattern.

> As soon as our venerable Chávez arrived at the place where the pueblo of Yuririapúndaro is found today, he endeavored to give the population the form of a republic and to this end he opened up streets, enlarged the plaza, established an ejido and everything else that a well organized community needs.[3]

To assure a water supply, he built a small reservoir at a spring near the convent from which water was conducted to the convent garden and the town.[4]

---

[1]Beaumont, IV, Book 2, Chap. 1, pp. 27-43. The early history of the region has been admirably analyzed by Jiménez Moreno and Gerhard (1972).

[2]Jiménez Moreno, p. 68, 75; Basalenque, pp. 127-28. Basalenque gives 1550 as the date of foundation of the town by Chávez, but Jiménez Moreno states that the Augustinian convent was founded in 1539.

[3]Escobar, p. 314.

[4]Basalenque, p. 126; Escobar, p. 311.

Chávez was not content with this modest hydraulic improvement. The land immediately north of the town lay below the level of the River Lerma and was swampy in the rainy season. He prevailed upon the Indians to build a deep ditch about two kilometers long from the Lerma to this low- lying area so as to create a large artificial lake, now known as Lake Yuriria, that served as a fishery for the village. The size of the lake fluctuated with the seasonal flow of the river, but, in modern times, it commonly covers an area about fourteen kilometers in length and four kilometers in width.[5] A firmly held colonial tradition affirmed that, after building the canal, the Augustinians and the Indian pueblo divided the lake between themselves: the Augustinians got the half lying east of the center of the convent, and the Indians got the western half as a reward for their labor. There is no direct documentary evidence of this compact, but it was treated as a generally accepted fact in later texts, and the landowners of the Valle de Santiago assumed its existence in their dealings with the Augustinians over the use of lake water in the eighteenth century.[6]

Until the early eighteenth century, Acámbaro continued to present the paradox of a settlement on a major river, dependent on the tenuous water supply of a small arroyo[7] (Figure 2.1). With the technology of the time it was difficult to build a permanent dam on a river as large as the Lerma, and the terrain did not always permit a direct diversion of river water.[8] Several haciendas in the valley made use of other arroyos to irrigate plots of wheat.[9] But in the first quarter of the eighteenth century, one begins to find clear evidence of irrigation from the River Lerma itself, and wheat production rose abruptly to a level that usually fluctuated around 12,000 fanegas.[10] A single hacienda, San Cristóbal, produced much of the wheat crop, and in a very exceptional year, 1771, it alone produced no less than 18,000 fanegas. We know that this hacienda possessed a dam on the Lerma; in 1780, the farmers in the Valle de Santiago, far downstream, felt threatened by the quantity of water that it diverted.[11] But since there were few formal grants or adjudications of water rights in the valley, details are lacking.

---

[5]Basalenque, p. 125; Escobar, pp. 310-11.

[6]AGN-T, Vol. 294, Exp. 1, f. 64; AGN-A, Vol. 97, Exp. 2, f. 10 et seq.

[7]See map accompanying relación geográfica of Yuririapúndaro; MNA.AH Colección Paso y Troncoso, leg. 33.

[8]AGN-M, Vol. 75, f. 179 (expression of impossibility of building masonry dam on river).

[9]El Obispado de Michoacán, p. 168. The following haciendas had fields of wheat that were probably irrigated from arroyos: San Juan de la Penitencia or Xaripeo (ACM, leg. 849, Exp. 1244); La Concepción and Santa Lugarda in 1727; Paraguaro in 1730 (AGN-T, Vol. 570, Exp. 1, f. 42). Possibly also: Milpillas in 1730 (AGN-T, Vol. 570, Exp. 1, f. 33v); San José de Apeo in 1707 (AGN-T, Vol. 232, Exp. 5, f. 5v).

[10]San Juan Rancho Viejo in 1778 (ACM, leg. 815, Exp. 283); Santa Catharina, c. 1730 (AGN-T, Vol. 570, Exp. 1, ff. 38-41); San Diego in 1756 (AGN-M, Vol. 75, f. 179). Wheat production: ACM, leg. 836.

[11]AGN-A, Vol. 97, Exp. 2, f. 10 et seq.

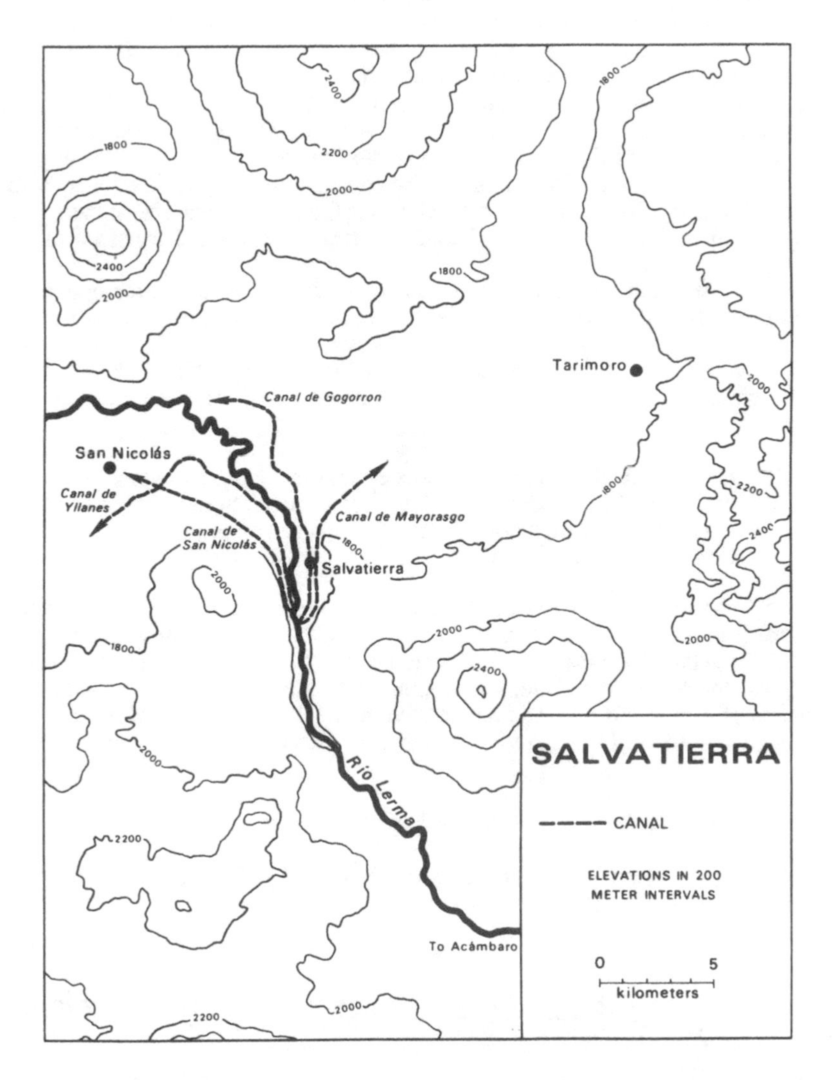

Figure 2.1 Sketch map of environs of Salvatierra

The water from the arroyo that supplied the original settlement of Acámbaro was carefully conserved and bitterly disputed during the colonial era. The water was conducted about six kilometers by a masonry conduit extending from a dam at the foot of the hills to the Franciscan convent and a public fountain in the town. Since the water was more pure than that of the River Lerma, it was highly valued for domestic uses. At some point, the Hacienda San Isidro acquired a right, later confirmed in the eighteenth century, to one-half of the water from the arroyo above the dam.[12] Farther upstream, the village of Tecuaro claimed a major share of the water from one of the springs feeding the arroyo, and the Haciendas Santa Clara and San Antonio customarily also diverted a portion of the flow.[13] Not surprisingly, all the parties came into conflict over the use of this modest water supply, and the pueblo of Acámbaro, the last in line, was heard to complain that the water was often dissipated before reaching the town.[14]

Two Indian pueblos near Acámbaro were the scene of similar conflicts. Tarandacuaro, which lay along an arroyo below a neighboring hacienda, received a formal grant of the excess flow (remaniente) in the arroyo but was forced to litigate the meaning of this concession. The pueblo of Emenguaro similarly claimed a right to the overflow (remaniente) from springs in a neighboring Carmelite hacienda and struggled to retain possession of four caballerías of land irrigated from a small canal on the River Lerma.[15]

Downstream from Acámbaro, the water works would be on a much grander scale and would enter more deeply into the social and economic history of the colonial period (Figure 2.2). On the plain lying between Lake Yuriria and the Lerma, the Augustinians formed a great estate, the Hacienda San Nicolás, during the sixteenth century. Basalenque writes that the Chichimec "general" Alonso de Sosa gave the land to the Augustinians.

> He gave much land in which was established an agricultural estate, called San Nicolás, and it grew so much, through the grace of God, that, among irrigated wheat haciendas, it took first place in New Spain.[16]

The "general" was probably a Pame or Guamare Indian leader who fought as an auxiliary to the Spanish around 1540 or 1550.[17] Although it is not clear how he could have transferred title to the vast expanse of the hacienda, the Augustinians did indeed use their influence over the Indian population to acquire land. An interesting document records nine transfers of land to the Augustinians in 1590 and 1592 by Indians who, for the most part, had just received land through viceregal grants. Part of this land lay along the River

---

[12]AGN-T, Vol. 534, 2a Parte, Exp. 1, Cuad. 1, f. 1035.

[13]AGN-T, Vol. 69, 1a Parte, Exp. 1, f. 32 and map; Vol. 866, Exp. 3, ff. 2, 150; Vol. 2716, Exp. 3, Cuad. 1, ff. 1-17.

[14]AGN-M, Vol. 74, f. 80.

[15]AGN-P, Vol. 23; AGN-M, Vol. 31, f. 238v; Vol. 32, f. 101v; AGN-T, Vol. 580, Exp. 1, Cuad. 1, ff. 20-26, 298, Cuad. 2, f. 184.

[16]Basalenque, p. 130.

[17]Powell, p. 166; Jiménez Moreno gives the name of the benefactor as Alonso de Castilla. Jiménez Moreno, p. 75.

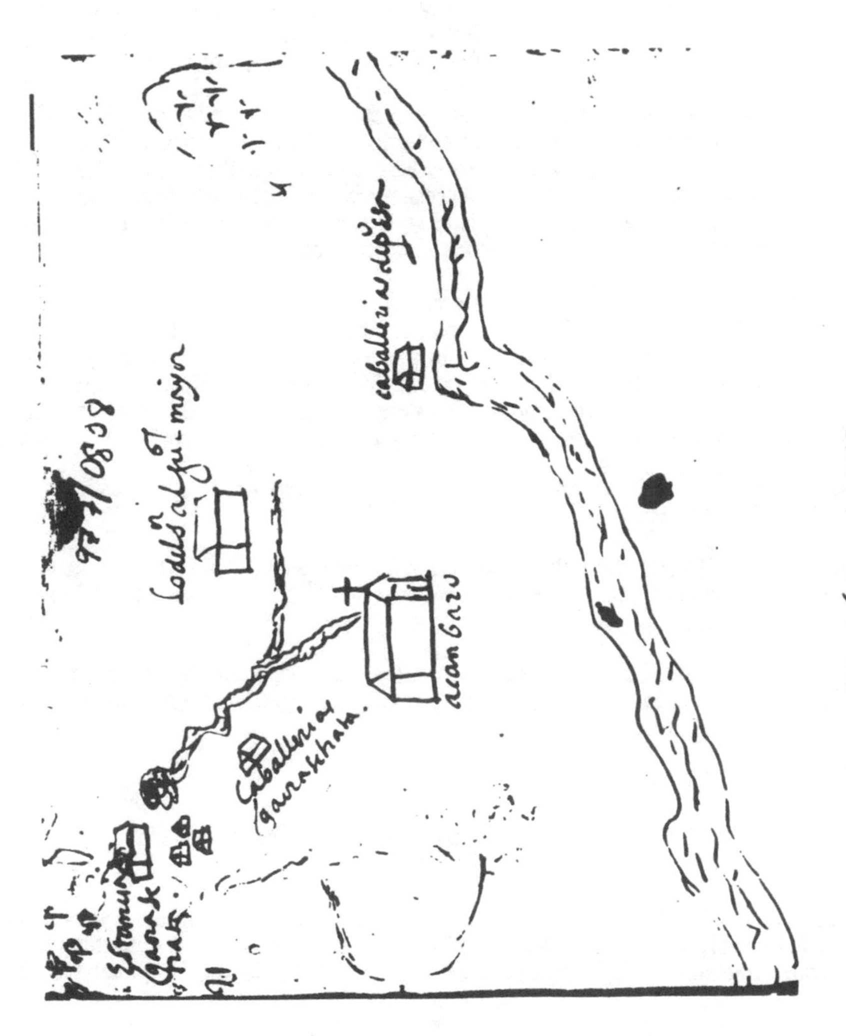

Figure 2.2 Acámbaro in 1615

Lerma.[18] The communal lands of an Indian pueblo, Taristarán, lying within the emerging domain of the hacienda, were also absorbed into the hacienda after a long legal battle.[19] But if the Augustinian archive in Mexico City were open to scholarly study, research might prove that such acquisitions were exceptional. The hacienda may well have expanded principally through purchase of neighboring lands. A large section of the hacienda was probably acquired from two Spaniards who had each received grants to twenty caballerías of land on the southwest bank of the Lerma in the sixteenth century.[20]

Sometime between 1580 and 1589, the Augustinians built a major canal about eight kilometers in length from a point above the falls of the Lerma to the Hacienda San Nicolás. The relación geográfica of Yuririapúndaro, written in 1580, states that no irrigation had been undertaken in the area although conditions were propitious.[21] A document dated 1589 describes a fully operational canal[22] (Figure 2.3). It was undoubtedly during this period that Fray Jerónimo de la Magadalena was put in charge of the Augustinians' lands near Yuririapúndaro. According to Basalenque,

> He undertook to create the wheat hacienda that we now call San Nicolás, and he assumed the administration of it; he built the dam, a very difficult feat, dug the canal many leagues in length that can irrigate . . . (land) from which one can harvest 50,000 fanegas. He built four flour mills and bought adjoining land so that he succeeded in forming a hacienda that was then valued at one hundred thousand pesos.[23]

The Augustinians appear to have secured a grant of water rights in 1606.[24]

As Basalenque asserted, the canal soon transformed the Hacienda San Nicolás into the largest wheat hacienda in New Spain. In the early seventeenth century, there were years when the hacienda produced 10,000 fanegas of wheat--more than the entire valley of Atlixco, the first center of wheat production near Puebla. It had 400 oxen, 150 mules, and 120 Indian laborers ("indios de racion, asi gananes como arrieros"). With annual

---

[18]C-AD, Fondo 507-3. The terms of the grants forbid conveyance to ecclesiastical bodies. AGN-M, Vol. 16, ff. 97 and 97v.

[19]AMS, June 7, 1717 and December 3, 1827; Navarette (1978), I, p. 190. Vicente Ruiz Arias has noted that land grant to the pueblo used in this litigation appears to be forged. See AGN-T, Vol. 1810, Exp. 14.

[20]AGN-M, Vol. 13, ff. 102, 109v. There were allusions to land acquired from Ponce de León and Juan Palacio in unclassified documents of the Augustinian Archive that I was permitted to review hastily.

[21]Corona Núñez, II, p. 67.

[22]AGN-T, Vol. 2809, Exp. 14, f. 5. See also references to canal in two 1590 land grants. AGN-M, Vol. 16, ff. 17v and 45v.

[23]Basalenque, p. 308. For a seventeenth century mention of the dam of the Augustinians, see BN-AF, caja 47, doc. 1059.1. A description of the hacienda in 1779 states that the canal traveled one and one-half leagues before reaching the hacienda. AA, unclassified document.

[24]See Appendix 1. I have not been able to find a transcript of this grant in the ramo de mercedes of the Archivo General de la Nación, but it is quite plausible to believe that it was made.

revenues in the order of 6,000 pesos, the hacienda was reasonably evaluated at 100,000.[25] Few estates, apart from sugar haciendas, had comparable worth.[26] The hacienda continued to prosper throughout the colonial era, although it lost some of its preeminence and never approached the level of wheat production foreseen by Basalenque.[27] In 1779, it had forty caballerías of land sown with wheat and three flour mills that operated the year around.[28]

Apparently sparked by the example of the Augustinians, a flock of Spanish settlers sought land grants on the west bank of the Lerma. Between 1583 and 1585, the audiencia issued no fewer than twenty-three grants of irrigable land, most expressly including water rights, and four licenses to construct flour mills.[29] Several of the applicants were from Celaya, where, as we have seen, there were cooperative efforts to build irrigation systems between 1567 and 1576.[30] But there is no evidence of any common organization here. Instead, an irrigation system was built sometime before 1589 by a single grantee with exceptional private resources, the same Juan de Yllanes who figures so prominently in the early history of Celaya. By the 1580s, his considerable wealth and long experience in irrigation enabled him to successfully undertake this new venture on the River Lerma. The head of his principal canal lay over the falls, just below the strategic location selected by the Augustinians; from this point, a large expanse of land along the west bank of the river could be easily irrigated. By 1590, he had built a second irrigation system downstream from the first.[31] Judging from later evidence, both systems obtained water from intakes cut into the river bank aided by small barriers jutting part way into the river.[32] Yllanes had received a grant of four caballerías of land and a license to build a flour mill and was associated, at least in the milling enterprise, with another large landowner, Lope de Sandi.[33] Over a period of twenty years, he and his heir, Martín Hernández, bought up most of the other grants one by one. Fragmentary records reveal that Yllanes acquired twenty caballerías before his death in 1590, and Hernández bought another nineteen caballerías, three sheep *estancias*, and various other

[25]*El Obispado de Michoacán*, p. 201; Basalenque, pp. 308, 330.

[26]Chevalier, p. 313.

[27]ACM, leg. 861.

[28]AA, unclassified doc.; Florescano and Gil Sánchez, p. 64.

[29]AGN-M, Vol. 11, f. 287v; Vol. 12, ff. 17v, 19, 22, 27v, 45, 46, 46v, 63, 63v, 150, 150v; Vol. 13, ff 28v, 62v, 64, 79, 97, 102, 109v, 110v, 162v; AGN-T, Vol. 580, Exp. 1, ff. 316, 317.

[30]E.g., Miguel Juan, Pedro Tellez de Fonseca, Bartholome de Entrambasaguas, Gregorio Gonzáles, Cristóbal Sánchez de Carabajal, Juan de Yllanez.

[31]AGN-M, Vol. 16, f. 117v; AGN-T, Vol. 2809, Exp. 14.

[32]AGN-T, Vol. 1247, Exp. 1, f. 35 (1974).

[33]AGN-M, Vol. 13, f. 28v; AGN-T, Vol. 2809, Exp. 14. Yllanez hald certain lands in common with Lope de Sandi. C-AD, fondo 507-2. For other acquisitions of Sandi, see AGN-T, Vol. 99, Exp. 1, f. 363v; AGN-M, Vol. 11, f. 287v.

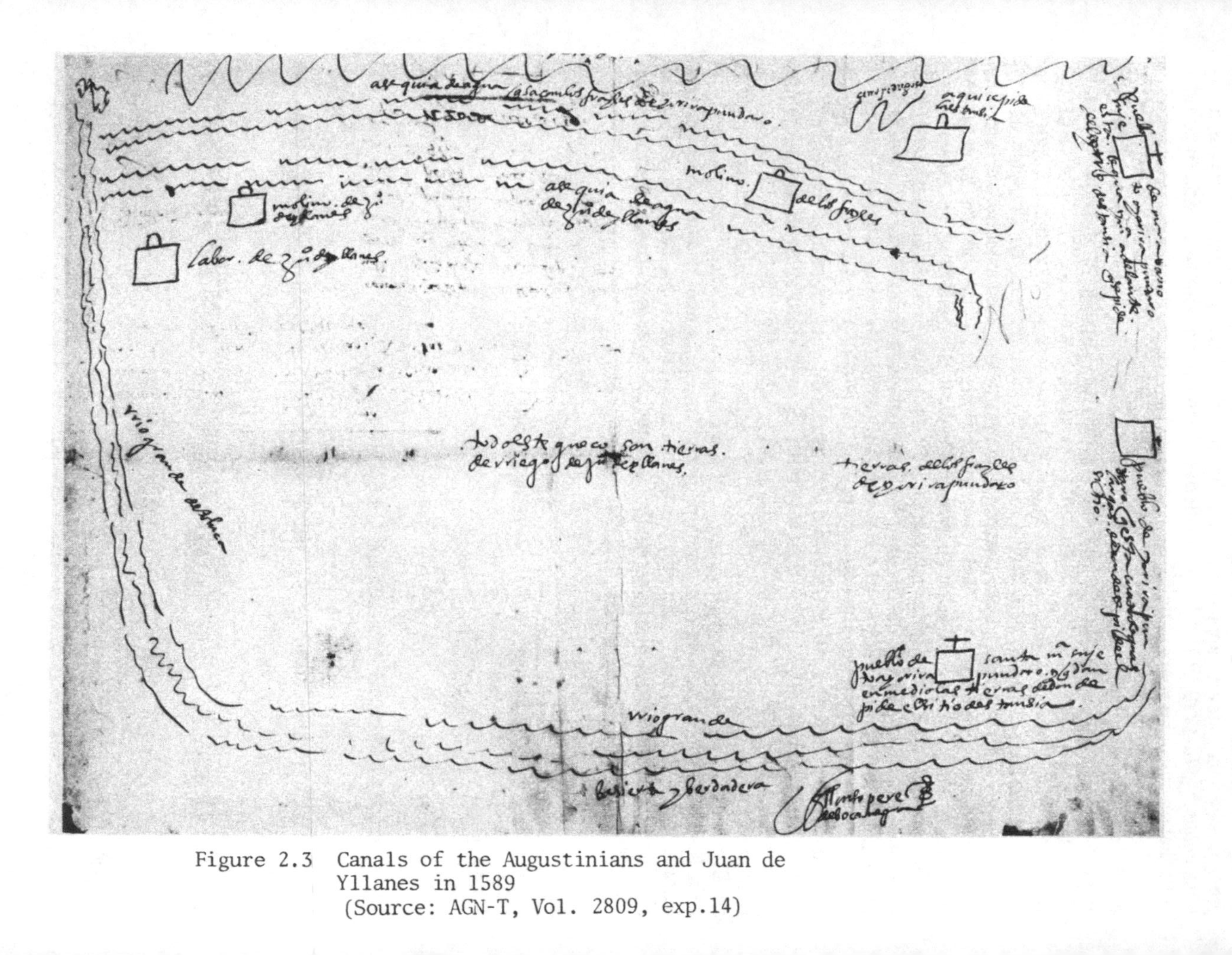

Figure 2.3 Canals of the Augustinians and Juan de Yllanes in 1589
(Source: AGN-T, Vol. 2809, exp.14)

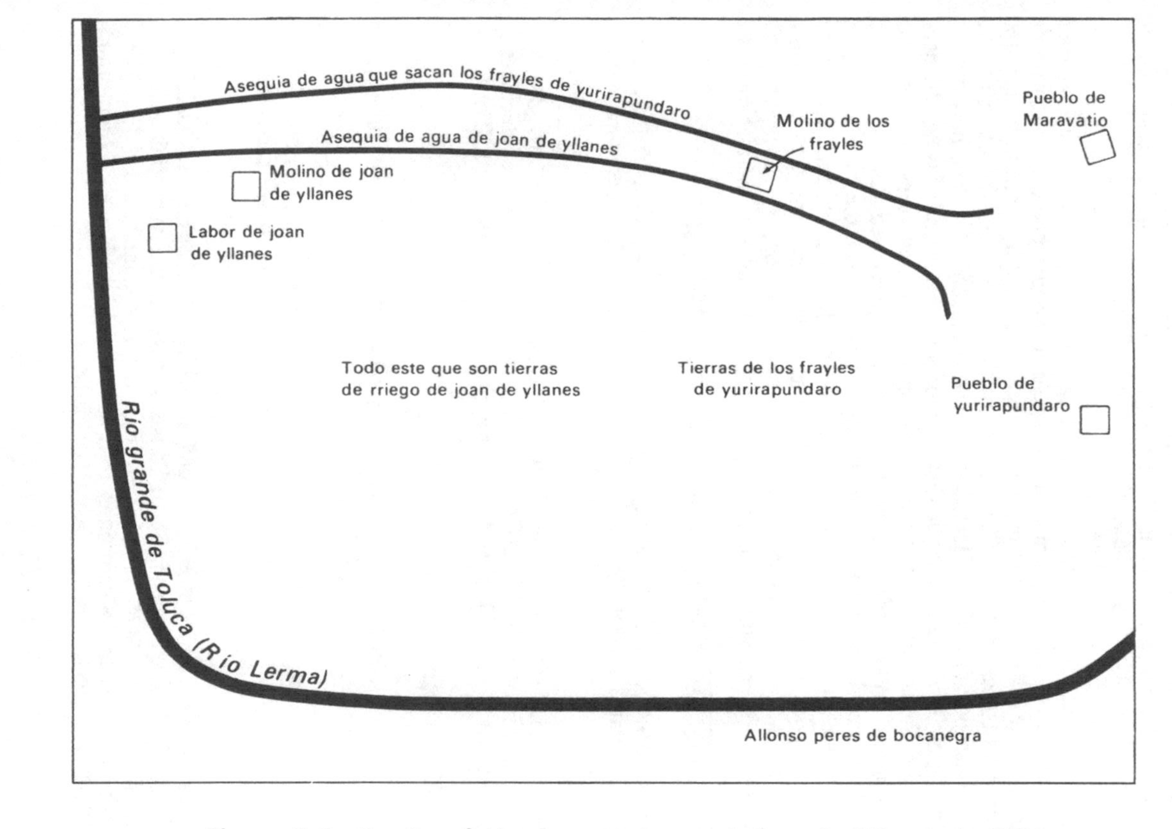

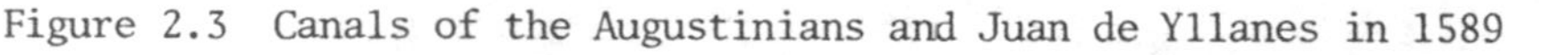
Figure 2.3 Canals of the Augustinians and Juan de Yllanes in 1589

properties.[34] These purchases, and probably many more, were consolidated in a single hacienda called San Buenaventura.

Upon the death of Martín Hernández in 1608, his estate was divided equally among his six children. On the west bank of the Lerma, the children each received land with rights to one-sixth of the water from the principal canal. This was considered enough for each to irrigate seven caballerías of wheat land. The Hacienda del Potrero was held jointly by all the heirs. The division of water was accomplished by the construction of masonry gates and irrigation boxes that effected a proportionate distribution of a continuous flow of water.[35] The smaller downstream canal may have fallen into the possession of one of the heirs, María de Torres, since it was later associated with her name.[36] The six heirs produced an aggregate quantity of wheat comparable to that of the Hacienda San Nicolás in the early seventeenth century.[37]

Around the middle of the seventeenth century, the property began to pass out of the hands of the Hernández heirs. The Convent of Santa Clara in Mexico City foreclosed on thirty caballerías of the land, which it soon sold to private parties;[38] Agustín de Carranza Salcedo, the royal scribe assigned to the area and a wealthy landowner, acquired the Hacienda del Potrero in 1644;[39] and the Carmelites and Augustinians came into possession of other portions of the Hernández estate before 1670.[40] The new owners continued the system of dividing a continuous flow of water by means of water dividers. An effort by the Carmelites to introduce a system of irrigation turns (tandas) was successfully opposed by a landowner who had recently purchased one of the other haciendas and had already built new masonry irrigation boxes. But the system operated fairly only if all the participants constructed and maintained facilities capable of effecting a proportionate division of the water. The Carmelites and another landowner independently sought judicial orders compelling other participants to repair their irrigation boxes.[41] By the eighteenth century, the shares of water of the Hacienda del

[34]AGN-M, Vol. 13, f. 28v; AGN-T, Vol. 99, f. 364v; C-AD, Fondo 507-2.

[35]C-AD, Fondo 507-2; AGN-T, Vol. 2951, Exp. 31; BN-AF, caja 47, doc. 1059.1 and 1059.4. I have translated the term "marco" as irrigation box or divider, according to the context.

[36]References to canal of María de Torres: MNA.AH, Eulalia Guzmán, leg. 107, doc. 7 (1726); AA, unclassified doc. (1670); AGN-M, Vol. 70, f. 101. Other vague references to second canal: AGN-T, Vol. 2951, Exp. 31; BN-AF, caja 47, doc. 1059.1 and 1059.4.

[37]El Obispado de Michoacán, pp. 165, 210.

[38]BN-AF, caja 47, doc. 1059.1. The hacienda seems to have remained in private hands. See AGN-T, Vol. 586, Cuad. 2, f. 6; AGN-P, Vol. 45.

[39]C-AD, Fondo 507-2 (sale of Hacienda Buenaventura to Martín Tamayo); BN-AF, caja 47, doc. 1059.4, f. 33 (sale of land with rights to two-sixths of water to Antonio Ramos Ramón).

[40]The land was still in the hands of heirs in 1647. AGN-T, Vol. 185, Exp. 1, Cuad. 2, f. 2; Vol. 2935, Exp. 109. The Hacienda Santo Tomás belonged to the Augustinians in 1696 and probably in 1670. AGN-T, Vol. 2791, Exp. 6, f. 1; AA, unclassified doc. The Carmelites held the Hacienda Concepción in 1667 and acquired the Hacienda Maravatio in 1665. AGN-T, Vol. 2951, Exp. 31; AA unclassified doc.

[41]AGN-T, Vol. 2951, Exp. 31; BN-AF, caja 47, doc. 1059.4.

Potrero had come to be expressed in terms of surcos, a unit of water measurement of the day, while the other haciendas claimed a right to fractional shares of the water flowing in the canal.[42]

The properties using the smaller downstream canal, known as the canal of María Torres, were acquired by the Augustinians and Carmelites, who agreed in 1670 to share the water equally. Each order built irrigation boxes at its own expense to divide the flow and assumed equal responsibility for maintenance. The transaction further increased the advantage of the Augustinians in the distribution of water. In dry years, the Carmelites were sometimes forced to ask the Augustinians for water from their canal.[43]

It was probably in the seventeenth century that the Carmelites extended the canal of Yllanes so that it crossed above the canal of San Nicolás by means of a short aqueduct and irrigated the Hacienda Maravatio to the south. The anomaly of crisscrossing canals has existed throughout modern times, and, if we assume that the names and locations of the haciendas have remained constant, it can be dated to the mid-seventeenth century.[44] But apart from this innovation and certain small, ephemeral irrigation intakes, the irrigation systems built on the west bank of the Lerma in the sixteenth century remained essentially unchanged throughout the colonial era. A document written in 1794 describes the dam of the Hacienda San Nicolás, the canal of Yllanes (then known as the canal of Maravatio), the smaller canal of María Torres (then leading to a fulling mill), as well as another irrigation ditch of minor importance.[45]

On the other side of the river, the bottomland extending eastward toward Apaseo, known as the Valley of Tarimoro, early attracted the attention of the encomendero of Acámbaro, Hernán Pérez de Bocanegra. In 1539, a year after he secured the encomienda, he obtained a grant of a cattle estancia in the valley. Other grants to his children followed, and the valley became a part of the extensive domain of the Bocanegra family.[46] But the heirs of Bocanegra encountered financial difficulties near the end of the century, and their property was auctioned off.[47] The purchaser of thirteen cattle estancias (each presumably no smaller than a sitio de ganado mayor or 1,755 hectares) was Gerónimo López de Peralta, royal treasurer of New Spain (tesorero de la real caja y hacienda) and regidor of Mexico City. The son of a conquistador who had received important posts and married a viceroy's niece, López displayed much business acumen and greatly increased his inherited wealth, acquiring, among other things, cattle estancias in what is now the southern part of the state of Guanajuato. He established three mayorazgos on his death in 1608. The first of these, given to his son Gabriel, included

[42]AGN-T, Vol. 586, Exp. 1, Cuad. 2, f. 65 (Hacienda del Potrero); ACM, leg. 849, Exp. 1237, f. 10 (Hacienda Concepción).

[43]MNA.AH, Eulalia Guzmán, leg. 107, doc. 7.

[44]The Hacienda Maravatio was irrigated in 1670 (BN-AF, caja 47, doc. 1059.4) and the canal itself was called the "acequia de Maravatio" in 1794 (AGN-T, Vol. 1247, Exp. 1, f. 35). It was earlier called the "acequia de los Parcioneros" (AGN-T, Vol. 586, Exp. 1, Cuad. 2, f. 65) or "acequia de los Hernández" (AGN-T, Vol. 185, Exp. 1, Cuad. 2, f. 2).

[45]AGN-T, Vol. 1247, Exp. 1, f. 35. Note that the canal of María Torres served a fulling mill. AGN-M, Vol. 70, f. 101.

[46]A summary of most of these grants is found in AGN-T, Vol. 99, Exp. 1, ff. 176-83v.

[47]López de Peralta seems to have purchased the estancias at an auction in 1579 (AGN-T, Vol. 99, f. 191v) although the purchase was evidenced by a decree dated 1605. Fernández del Castillo, p. 268.

the estancias acquired from the Bocanegra heirs, thirty-four other estancias in the region, and many more properties. It was valued at the extraordinary sum of 294,000 pesos.[48]

Another of the richest men in New Spain, Pedro de Arismendi Gogorrón, was also acquiring estancias along the east bank of the Lerma at the beginning of the seventeenth century. One of the first settlers in San Luis Potosí, he had gained a large fortune in the mines, which he invested in further mining ventures, a great northern hacienda, and other enterprises. Chevalier regards him as a prototype of the modern entrepreneur engaged in multiple enterprises, all on a large scale, with investments of 50-, 60-, or 80,000 pesos.[49] In 1615, he appears as the owner of ten estancias on the east bank of the Lerma within the jurisdiction of Yuririapúndaro. The southernmost extension of these properties border on the estate of López de Peralta in the Valley of Tarimoro.[50]

Agriculture came late to this area as it did to much of the Bajío. While the cattle estancias of López de Peralta produced some maize in the early seventeenth century, those of Gogorrón appear to have been largely empty.[51] But, faced with declining herds of cattle, both men undertook independently in 1618 to convert a portion of their properties into irrigated wheat haciendas.[52] Relying on a license to build a flour mill, López de Peralta built a dam above the falls on the River Lerma. The water was channeled to the northeast by way of ditches and masonry aqueducts to the flour mill and then to fields on the valley floor. López claimed to have spent 20,000 pesos on the project.[53] He later secured a more ample grant of water rights in 1631.[54] For his part, Gogorrón built a dam along the river closer to the falls under the authority of a formal grant of water rights. It similarly channeled water north through ditches and masonry conduits to land on the east bank of the river. Any excess water drained into the arroyo of Tarimoro about twelve kilometers from the point of origin. To improve his claim of title, Gogorrón secured a grant of fifteen caballerías of land within the new irrigation system.[55]

[48]Chevalier (1952), pp. 24, 53, 130, 189; Fernández del Castillo, p. 267; Fernández de Recas, p. 76 et seq.

[49]Chevalier (1952), pp. 219, 230.

[50]AGN-M, Vol. 30, f. 42v. The cerro of Culiacán mentioned in this document forms the southern boundary of the valley of Tarimoro.

[51]See map accompanying relación geográfica of Celaya. MNA.AH, Collección Paso y Troncoso, leg. 33.

[52]On the phenomenon of declining cattle population, see Chevalier (1952), pp. 134-37; AGN-R, Vol. 8, Exp. 1, f. 1 (complaint of López de Peralta).

[53]AGN-M, Vol. 34, f. 43 (May 16, 1618); AGN-R, Vol. 8, Exp. 1, f. 1; AGN-T, Vol. 185, Exp. 1, f. 6v; Vol. 988, Exp. 1, Cuad. 4, f. 41.

[54]AGN-T, Vol. 822, Exp. 2, Cuad. 1, f. 1.

[55]AGN-M, Vol. 33, f. 258v (July 7, 1618); Vol. 35, f. 40v; AGN-T, Vol. 1247, Exp. 1, f. 6.

Gogorrón's land was rented to several farmers and evidently produced on the order of 6,000 fanegas of wheat in the early seventeenth century.[56] It passed to his son, and, upon the son's death, it was divided equally among five heirs. The heirs entered into a formal written agreement in 1646 to resolve their respective rights to water. They stipulated that (1)they could each draw up to one-fifth of the water from the canal for the irrigation of their haciendas; (2)they would each be free to invest, at their own expense, in improving the dam and increasing the capacity of the canal so that water could be conducted to new irrigated lands; and, (3)they would not sell the water remaining after irrigation of their fields but would return it to the canal (Figure 2.4).

The properties of López de Peralta's mayorazgo had a much more complicated history. Gabriel López de Peralta, who seems to have lacked his father's financial ability, needed to circumvent the restriction on alienation of mayorazgo land to obtain funds. He subrogated revenues of the land, or conveyed it by perpetual lease (*a censo perpetual*), perhaps followed by a relinquishment of the right to further payments, or simply sold it outright without any pretense of legality. In this way, he disposed of several parcels of land, often with associated water rights, along the lower reaches of the canal.[57] But the purchasers of water rights received access to a very tenuous supply and were forced in dry years to borrow water from the canal of Gogorrón.[58] The flow of the canal was evidently limited by poor design of the system and declined with inadequate maintenance; around 1631, wheat production on the mayorazgo land was similar to that of the estate of Gogorrón,[59] but in 1646 the irrigation system was capable of irrigating only eight caballerías because of the poor construction of the dam and the inadequate capacity of the canal.[60]

Apart from the parcelization of the mayorazgo estate, the small creole population of the valley was hemmed in on all sides by latifundia.[61] The men of Spanish descent had only one realistic possibility of acquiring land: the foundation of a city that might grant land to its vecinos. As Chevalier observes, in the struggle for land the situation of the small creole farmer who did not belong to a villa or a *ciudad* was "almost with hope."[62]

Throughout the first half of the seventeenth century, there were persistent efforts to found a city along this stretch of the River Lerma. In 1603, a group of Spaniards sought unsuccessfully to interest the Augustinians in the foundation of a city on the

[56]*El Obispado de Michoacán*, p. 165. The text dated 1631 contains alternative figures in the form of interlineations and sometimes seems to indicate excessively high production. The tithes collectors had an incentive to exaggerate the quantities produced. Compare: ACM, leg. 861.

[57]AGN-R, Vol. 8, f. 1; AGN-T, Vol. 351, Cuad. 2, ff. 8v-20, 122; AMS, year 1652.

[58]AGN-T, Vol. 185, Exp. 1, Cuad. 1, ff. 6v-20, Cuad. 2, f. 12v, Vol. 988, Exp. 1, Cuad. 4, f. 41.

[59]*El Obispado de Michoacán*, p. 165.

[60]AGN-T, Vol. 185, Exp. 1, Cuad. 1, f. 15.

[61]In the petition of Carranza Salcedo for the foundation of the town, forty persons offered to become vecinos. Some 139 persons signed petitions concerning municipal offices. Vera identifies ten additional persons as vecinos who had houses in the valley. Vera, p. 45.

[62]Chevalier (1952), p. 297.

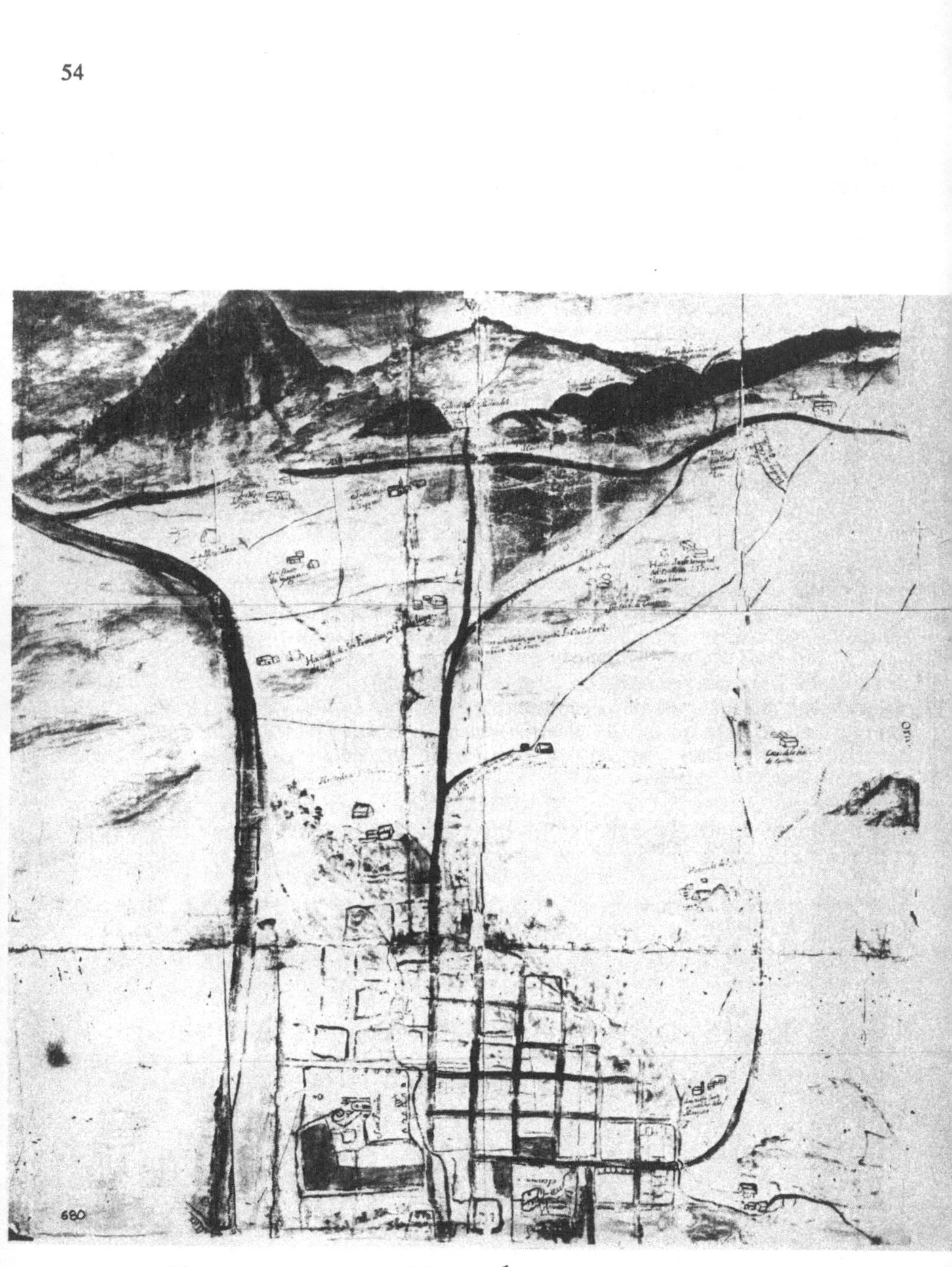

Figure 2.4 Canals of Gogorrón and the Mayorazgo, Salvatierra, 1724 (Source: AGN-T, Vol. 351)

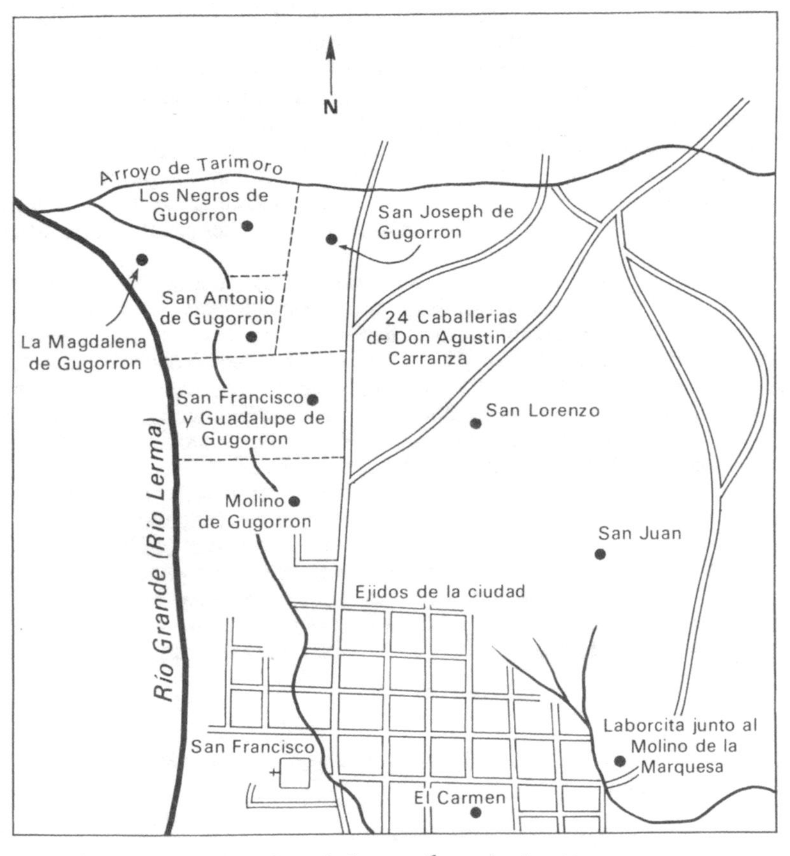

Figure 2.4 Canals of Gogorrón and the Mayorazgo, Salvatierra, 1724. Interpretation of original map dated 1724.

Hacienda San Nicolás.[63] Soon afterwards, Martín Hernández sponsored a petition to establish a town on the east bank, but it was rejected in 1610 as a result of the opposition of Gogorrón, who stood to lose in the distribution of land.[64] Agustín Carranza de Salcedo began a further effort to secure viceregal authorization in 1632 that was pursued over many years in petitions that disclose plans for the distribution of vecindades of "land and water."[65]

These efforts finally bore fruit in 1644. The Spanish residents reached an agreement with Gabriel López de Peralta: in exchange for the donation of land for the city, he and his descendants would be given certain prestigious titles and municipal offices, six house sites (solares) within the city, and an annual annuity of 2,000 persos to be paid directly out of the royal alcabalas of the city. In separate petitions, López de Peralta and Carranza Salcedo put the proposal before the audiencia. Both extolled the valley and its irrigation works, which produced wheat for "Mexico City, San Luis Potosí, Zacatecas, and other areas." López de Peralta proposed that his land be divided into vecindades consisting of four caballerías of land and eight days of water, but he expressly excluded from his offer fifteen caballerías of irrigated land on the upper portion of the canal and the water needed for their irrigation ("con el agua necesaria para el riego de ella, sin que para el uso de la dicha agua haya de entrar en tandas").[66] Carranza Salcedo's petition was largely devoted to the vital matter of the sale of municipal offices, a source of revenue to the crown, but it repeated the proposal to create vecindades with eight days of water from the canal of the mayorazgo or another canal that might later be built.[67]

The establishment of the city of San Andrés de Salvatierra was authorized by a viceregal order dated February 9, 1644. Each of the original vecinos would be given four caballerías of land with appurtenant water rights ("para riego de ellas, el agua necesaria de la del Rio Grande o acequias"), as well as two house sites (solares) and four plots (suertes) for gardens, vineyard and olive orchard. The vecinos were granted permission to build jointly at their own cost an additional irrigation works along the river. Vecinos who should fail to pay their proportionate share of the cost of such improvements could forfeit their municipal citizenship together with their vecindad.[68]

On January 3, 1646, a judge, Pedro de Navia, was commissioned to effect the distribution of land. His order of appointment again provided that every vecindad would have a right to a certain number of days of water. Upon the arrival of Navia in the valley, it developed that López de Peralta had a trick up his sleeve. He had described the offered land vaguely; he now sought to donate land that was also claimed by an heir of Martín Hernández. He evidently hoped to gain the rewards of founding the city and dispose of a troublesome lawsuit in one stroke.[69] Navia also discovered that the irrigation system of

---

[63]Basalenque, p. 330.

[64]AGN-T, Vol. 988, Exp. 1, Cuad. 4, ff. 43-5. Vera says that Arismendi Gogorrón defeated the plan. Vera, p. 40.

[65]AGN-T, Vol. 98, Exp. 1, Cuad. 4, f. 39v; Vera, p. 41.

[66]Báez Macías, pp. 671-92; AGN-T, Vol. 185, Exp. 1, f. 6; Vol. 988, Exp. 1, Cuad. 4, ff. 39v-41v.

[67]Vera, p. 46.

[68]AGN-T, Vol. 988, Exp. 1, Cuad. 4, ff. 45-48; Vera, p. 61.

[69]Báez Macías, pp. 672, 692; AGN-T, Vol. 185, Exp. 1, Cuad. 1, ff. 65-66v.

the mayorazgo would not suffice to irrigate new lands and that the construction of a new system would be costly. ("No son faciles de hacer los tomas y presas del agua del rio grande para su riego y beneficio como esta mandado por clausa expreso de la comision.")[70]

The first distribution of land to the principal office holders of the city on April 13, 1646 did not mention water rights, and the land lay to the east of the mayorazgo in an area not easily accessible by irrigation canals.[71] But on September 6, 1646, an expert was appointed to study the feasibility of an extended canal:

> How many caballerías of land can be irrigated, what it would cost to build the intake, dam and canal . . . so that in the future each beneficiary may be made responsible for the cost of the canal.[72]

In October, the expert submitted his reports: much of the valley (150 caballerías) was suitable for irrigation. However, the dam of the mayorazgo would have to be enlarged, the initial portion of the canal widened, and a major branch canal built a distance of 700 varas over uneven ground, requiring the construction of aqueducts and ditches lined with masonry to prevent water loss. The expert modestly estimated the cost of these improvements at 20,000 pesos.[73]

For several years, the cabildo and commissioned judges continued to parcel out mayorazgo land to the vecinos of the new city.[74] Three more judges were commissioned for this purpose as well as to adjudicate disputes with López de Peralta and other matters. But López himself received nothing in his lifetime in compensation for what he had lost. The cabildo of Salvatierra, which felt deceived by his offer of land, refused to grant him the promised offices and annuity.[75] It was not until 1707, after seventy-two years of litigation, that his heirs were finally compensated.[76]

The plan of irrigating the new vecindades did not die immediately. In 1653, one still finds a reference to "the canal that is to come from the Río Grande."[77] But none of the subsequent distributions of land mention water rights, and nothing was done. The revenues from the rental of town lands were inadequate for such a project, and they were soon pledged to the Carmelites to finance the construction of a bridge across the River Lerma.[78] Moreover, the vecinos no longer had an economic incentive to finance the

---

[70]AGN-T, Vol. 185, Exp. 1, Cuad. 1, f. 1.

[71]Báez Macías, p. 695; AGN-T, Vol. 185, Exp. 1, Cuad. 2, f. 36.

[72]AGN-T, Vol. 185, Cuad. 1, f. 9.

[73]Ibid, Cuad. 1, f. 15, Cuad. 2, f. 14.

[74]AGN, Templos y Conventos, Vol. 24, Exp. 2, f. 35; AGN-T, Vol. 879, Exp. 8; Vol. 1446, Exp. 2, f. 5 (location of first vecindades); AGN-M, Vol. 49, f. 51; Vol. 53, f. 70.

[75]Vera, p. 71-72; AGN-T, Vol. 185, Exp. 1, Cuad. 1, f. 45.

[76]AGN-T, Vol. 2816, Exp. 1.

[77]AGN-T, Vol. 185, Exp. 1, Cuad. 1, f. 58. See also AGN-M, Vol. 56, f. 24.

[78]Báez Macías, p. 682. See also AGN-A, Vol. 108, Exp. 1, 7, and 12 (revenues in late eighteenth century).

project collectively as contemplated by the viceroyal order. The economy of the northern mining center, the basis for the agricultural prosperity of the Bajío, had collapsed. Silver production in Zacatecas declined catastrophically in the late 1630s and did not recover for twenty-five years. Other mining centers had experienced an earlier decline.[79] A perceptive contemporary, Fray Diego de Basalenque, writing around 1644, attributed the poverty of the region to an oversupply of wheat, and he expressed the fear that the establishment of the new city of Salvatierra would aggravate the situation.[80] The time, in short, was not propitious for new investment in irrigation works.

The offer of López de Peralta to include water rights in the donation of his land to the city did have at least a technical legal consequence. In 1652, the corregidor of Salvatierra and the city's regidores formally took possession of the canal of the mayorazgo in the name of the city, claiming the water was needed for houses and gardens. This act of possession was ratified within the year by an order of the audiencia.[81] The water from the canal was indeed diverted to land within the town, but the municipality never assumed responsibility to maintain the canal or sought to regulate its use. The canal continued to be known as the "canal of the mayorazgo."[82]

Despite its official rank as a city (ciudad), Salvatierra retained a rural character in the eighteenth century. A meticulous map prepared in 1758 shows that within the grid pattern of streets and lots most of the land was devoted to gardens. There were no notable buildings and little urban development of any kind. Small houses were found in open lots surrounded by trees. To the south of the city along the canal of the mayorazgo lay the Indian barrio of San Juan, similarly surrounded by small agricultural plots. A description of the city in 1761 reveals that water from the canal of the mayorazgo was channeled to these gardens and plots within the grid of streets (planta) and to the barrio San Juan (Figure 2.5).[83]

The portion of the mayorazgo estate in the valley of Tarimoro that survived the distribution of vecindades consisted of two moderately sized properties, the Haciendas San Lorenzo and San Juan, immediately adjoining the town.[84] To the north and east, other haciendas of nearly comparable size owed their existence to the voluntary sale and forced alienation of mayorazgo land.[85] The Carmelites, who arrived in Salvatierra in 1644, emerged as the most important landowners. They had received grants of extensive properties of the northern edge of the city to help finance the construction of the convent

[79]Bakewell, pp. 61, 220.

[80]Basalenque, p. 314.

[81]AGN-M, Vol. 49, f. 50; AGN-T, Vol. 988, Exp. 1, Cuad. 4, f. 50.

[82]AGN-A, Vol. 108, Exp. 5, f. 22. The name appears in the municipal appropriation of money to repair the canal in 1776--the only record of this sort of expenditure I have found.

[83]AGN-T, Vol. 535, Exp. 4, ff. 12, 58 (foundation of barrio San Juan); Vol. 822, Exp. 1, f. 197; Vol. 988, Exp. 1, ff. 103-4.

[84]Báez Macías, p. 709.

[85]For example, Hacienda de los Coyotes, eighteen caballerías in 1723, (AGN-T, Vol. 1446, Exp. 2, Cuad. 2, f. 5); Hacienda de los Panales, eleven caballerías in 1787 (ACM, leg. 852, Exp. 1254).

church and the bridge over the Lerma.[86] In 1660, they purchased the Hacienda San José from one of the five heirs of the Gogorrón estate. They soon sold the hacienda but repurchased it later in the century[87] and acquired another part of the Gogorrón estate, the Hacienda San Antonio, and former portions of the mayorazgo estate.[88] Their presence around Salvatierra in the mid-eighteenth century was so dominant that Báez Macías has said that the town "was close to becoming a monastery town or a Carmelite feifdom."[89] Although the Carmelites sold part of their holdings on the west bank, the order held two very large haciendas on each side of the river, together with much urban property, at the close of the century.[90]

The later colonial history of the irrigation system of the mayorazgo was marked by decline and contraction. The canal ceased to reach lower properties beyond the border of the mayorazgo in the late seventeenth century. There is no record of when these lands returned to dry farming, but in the eighteenth century, the mayorazgo itself experienced shortage of water. The progressive deterioration of the system brought the mayorazgo into conflict with small properties near the head of the canal, including the Hacienda Esperanza, an acquisition of Gabriel López de Peralta, with about two caballerías of irrigated land that passed into other hands but continued to be irrigated from the canal. In 1775, the owner of the hacienda fell into litigation with the mayorazgo over water rights.[91] More serious were the conflicts between the mayorazgo and the Carmelites. In 1692, after a period of litigation, the Carmelites secured an agreement from the heir of the estate giving them the right to irrigate a plot next to the convent known as the quadrilla baja. The settlement did not avoid a further dispute around 1720.[92] Still later in the century, the Carmelites dug a ditch from the canal of the mayorazgo to irrigate another adjoining plot of land, one caballería in size, known as the quadrilla alta. The volume of water in the canal was so reduced by this date that this modest diversion almost exhausted the flow so that it could not power the mill at the border of the mayorazgo estate. Portracted litigation followed, but there is no record of how it ended (Figures 2.4 and 2.5).[93]

---

[86]A meticulous account of these acquisitions is found in Ruiz Arias (1976), pp. 29-32. See also AGN, Templos y Conventos, Vol. 24, Exp. 2; AGN-M, Vol. 56, f. 24.

[87]AGN-T, Vol. 380, Exp. 4, f. 74; Ruiz Arias (1980); AMS, November 7, 1703, ff. 38-48.

[88]AGN, Vinculos, Vol. 166, f. 156v; AGN-T, Vol. 1247, Exp. 1, Cuad. 1, f. 6 (the Haciendas San Antonio, San José, and Guadalupe were formerly part of the Gogorrón estate).

[89]Báez Macías, p. 683.

[90]AGN-P, Vol. 45, Cuad. 1, ff. 159, 169; MNA.AH Eulalia Guzmán, leg. 104, pt. 1, doc. 11.

[91]AGN-T, Vol. 822, Exp. 2, Cuad. 1, ff. 62v-94v, Cuad. 2, f. 25; Vol. 988, Exp. 1, Cuad. 2, ff. 3v-40.

[92]AGN-M, Vol. 56, f. 24; AGN-T, Vol. 822, Exp. 1, Cuad. 1, f. 1, Cuad. 2, ff. 89-104; Ruiz Arias (1976), p. 30.

[93]AGN-T, Vol. 822, Exp. 2, Cuad. 2, f. 46; Vol. 988, Exp. 1, Cuad. 2, f. 144v, Cuad. 4, f. 120v.

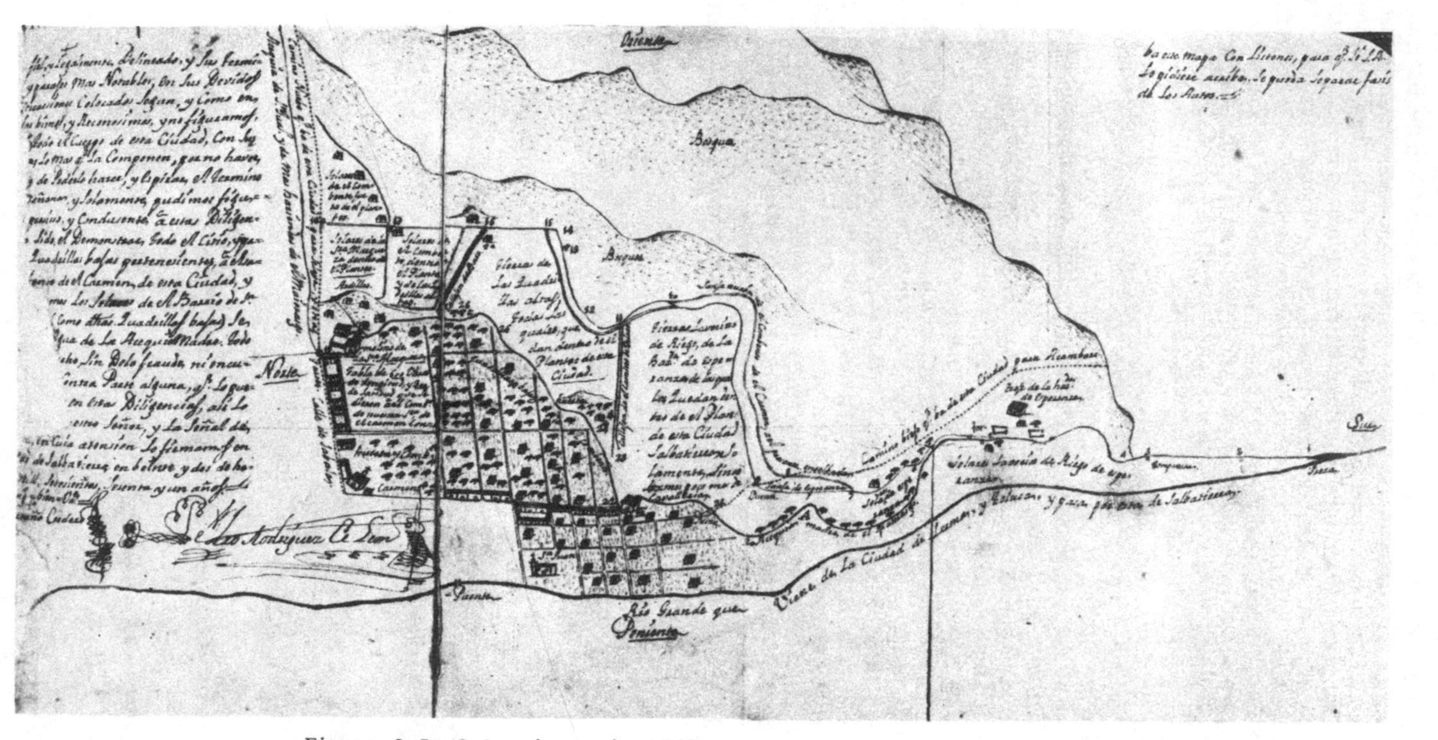

Figure 2.5 Salvatierra in 1761
(Source: AGN-T, Vol. 812, exp.1)

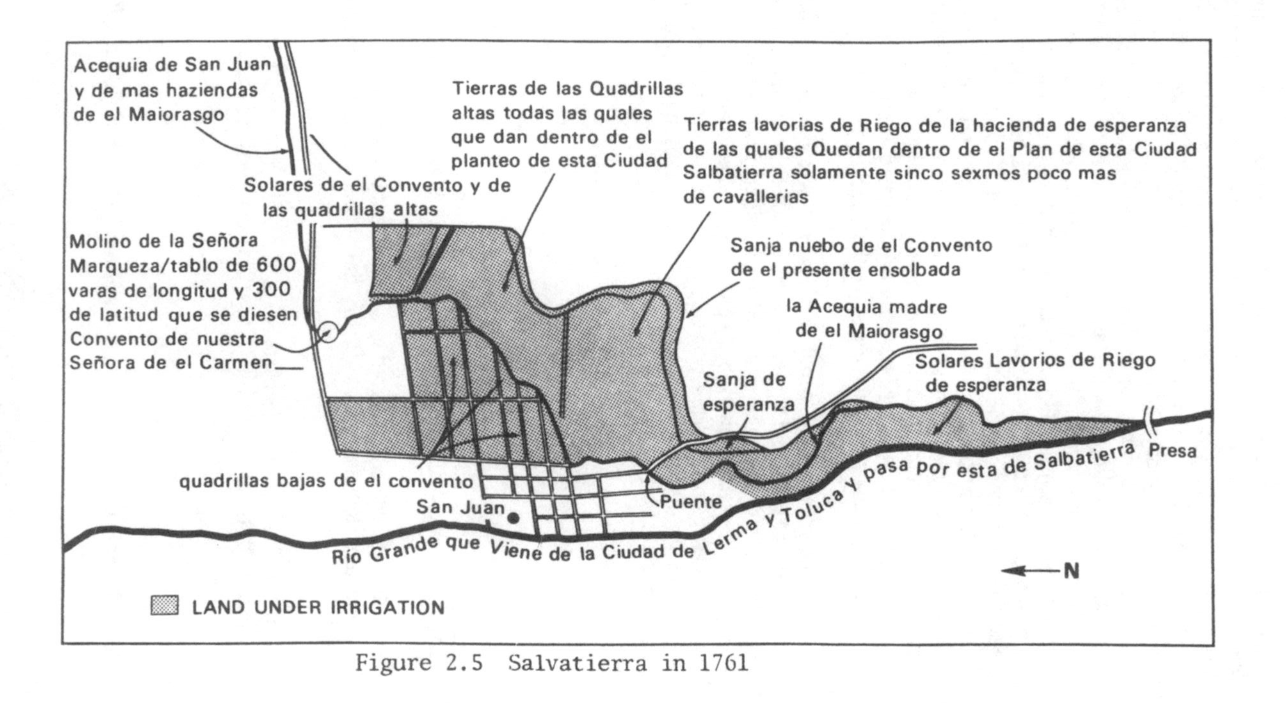

Acequia de San Juan y de mas haziendas de el Maiorasgo
Tierras de las Quadrillas altas todas las quales que dan dentro de el planteo de esta Ciudad
Tierras lavorias de Riego de la hacienda de esperanza de las quales Quedan dentro de el Plan de esta Ciudad Salbatierra solamente sinco sexmos poco mas de cavallerias
Solares de el Convento y de las quadrillas altas
Molino de la Señora Marqueza/tablo de 600 varas de longitud y 300 de latitud que se diesen Convento de nuestra Señora de el Carmen
Sanja nuebo de el Convento de el presente ensolbada
la Acequia madre de el Maiorasgo
Sanja de esperanza
Solares Lavorios de Riego de esperanza
quadrillas bajas de el convento
San Juan
Puente
Presa
Río Grande que Viene de la Ciudad de Lerma y Toluca y pasa por esta de Salbatierra
N
LAND UNDER IRRIGATION

Figure 2.5 Salvatierra in 1761

These conflicts were clearly a consequence of the poor maintenance of the canal. At mid-century, the dam of López de Peralta was no more than a pile of loose stones and dirt; thirty yards from the dam much of the water escaped to the river through a breach in the canal; downstream, the canal lacked properly constructed irrigation gates and outlets that could control the distribution of water. The mayorazgo relied on the Indians of the barrio San Juan to clean the upper portion of the canal in exchange for their use of water.[94] But the shortcomings of the irrigation system that provoked such chronic litigation did not lead to investment in the physical construction of the dam or to cooperative endeavors to improve it. On the hacienda itself, however, there is some evidence of hydraulic improvements. In the late eighteenth century, if not earlier, the hacienda possessed a system of storage basins.[95]

The canal of Gogorrón experienced a better fate. The haciendas along the entire length of the canal remained under irrigation,[96] and, during much of the eighteenth century, they each retained the original one-fifth share of the water. When the Carmelites constructed a new water divider (torno) in 1735 that had the effect of diverting more than their share, the action provoked an armed confrontation of hacienda administrators and led to an order of the audiencia restoring the traditional division.[97] Later documents, which speak only of "the water that belongs to" the haciendas, clearly refer to the shares based on the early division of the flow among Gogorrón's heirs.[98] The Carmelites evidently also built storage facilities (depósitos) to allow more extensive irrigation of their properties. These improvements are first described in an article dated 1865, but it seems likely that they were built during the late colonial era since that was the period of greatest prosperity for the order.[99]

Like the canal of the mayorazgo, the canal of Gogorrón was tapped by houses, gardens, and workshops within the town and served as a kind of de facto municipal canal. But once a year, the Carmelites and the owners of the other irrigated properties cleaned the canal and closed all outlets to assert their right to possession. The owner of a textile workshop tapped the canal undetected for many years by means of an underground pipe, but the Carmelites discovered the theft, closed the pipe, and, in subsequent litigation, soundly defeated his claim of right based on past usage.[100]

---

[94]AGN-T, Vol. 822, Exp. 2, f. 105; Vol. 988, Exp. 1, Cuad. 4, ff. 86v-87, 121.

[95]Swan (1977), p. 174. Although we have referred to a single "hacienda of the mayorazgo," Swan followed later documents that identify the property as the Haciendas San Juan and San Lorenzo.

[96]AGN-T, Vol. 435, Exp. 1, f. 31; Vol. 627, 1a parte, Exp. 1, f. 25v; Vol. 630, Exp. 1, f. 56; Vol. 1247, Exp. 1, Cuad. 1, ff. 6, 324; Vol. 1403, Exp. 2, f. 11v; Vol. 1816, Exp. 1, f. 248.

[97]AGN-C, Vol. 359, Exp. 5.

[98]MNA.AM-Qro, rollo 5 (Antonio García Botello, 18 Agosto 1710); AMS, 7 Mar 1703, f. 38-48 (1703).

[99]Vera Quintana, p. 585.

[100]AGN-T, Vol. 1247, Exp. 1, Cuad. 1, e.g., ff. 33-76, 117v, 298-99.

The colonial history of irrigation in Salvatierra was thus shaped by four latifundia founded in the late sixteenth and early seventeenth centuries. Cooperative endeavors in irrigation never took hold here; the city itself, despite ambitious early plans, assumed a somewhat parasitical relation toward the two canals that traversed it. But some new investment is apparent in the construction of storage basins and the extension of the original canal of Yllanes, and the four irrigation systems acquired a degree of complexity through private agreements and the division of water rights among heirs.

3
Valle de Santiago

The central Bajío, north of Salvatierra, where the Lerma curves around the small plain known as the Valle de Santiago, was progressively colonized by cattle estancias and dispersed farming settlements in the latter half of the sixteenth century. The indigenous population was small.[1] In the hills immediately to the south of the Valle de Santiago, there were two Indian pueblos, Araceo and Parangueo, but the flat land was the domain of the nomadic Chichimec tribes.[2] Despite harassment by Indian raids, Spaniards successfully established cattle estancias in the region during the mid-sixteenth century. The cattle population flourished to a phenomenal degree on the virgin grassland. The early maps of Celaya, Yuririapúndaro, and San Miguel in the 1580s depict plains populated in every direction by cattle.[3] In the opinion of Jiménez Moreno, the region of the Valle de Santiago was already "intensely populated" around 1570.[4] Certainly, the earliest description of the valley and adjoining areas north of the Lerma at the beginning of the seventeenth century reveal a landscape already filled with cattle estancias. Although many can be traced to viceregal grants,[5] the boundaries of the estancias were always vague and fluid.

The need to supply the real de minas of Guanajuato, founded in 1557, had led to the early establishment of farming settlements in nearby areas. Irapuato was formally organized as a congregación in 1589.[6] A few miles away on the northern bank of the Lerma, there may also have been some incipient Spanish agriculture. In 1577, one Juan Nuñez de Xerez received a license to build a flour mill on the Lerma near its junction with the River Laja.[7] In all likelihood there was some agricultural production, including irrigated wheat farming, within the Valle de Santiago itself. We know that at the beginning of the seventeenth century there was an irrigation canal and drainage ditch

---

[1]Gerhard (1972), pp. 121-23.

[2]AGN-T, Vol. 294, Exp. 1, f. 109.

[3]Jiménez Moreno, p. 90; Chevalier (1952), pp. 114-20; MNA.AH Colección Paso y Troncoso, leg. 33.

[4]Jiménez Moreno, p. 85.

[5]E.g., AGN-T, Vol. 118, Exp. 1, Cuad. 1, ff. 63-65, Cuad. 3, ff. 3-4v; Vol. 294, Exp. 1, f. 103; Vol. 311, Exp. 2, f. 14; AGN-M. Vol. 8, f. 25v; Vol. 10, f. 200v.

[6]Jiménez Moreno, ff. 76-7; AGN-T, Vol. 118, Exp. 1, Cuad. 3, f. 9 (mention of labradores of Irapuato in 1573).

[7]AGN-M, Vol. 10. f. 200v.

(desaguadero) along a flood channel of the Lerma, known as the Brazo de Moreno, where water could be easily diverted to fields. The "acequia del Brazo de Moreno," first mentioned in a document dated 1606, may well have been in existence for many years.[8]

The continued prosperity of the northern mines around 1600 favored the production of wheat in the central Bajío. At the same time, the decline in the number of livestock, evidently due to an ecological cause, clouded the future of cattle ranching. The Valle de Santiago had good land and access to water; the neighboring land on the northern bank of the Lerma was believed to have similar qualities. But the settlers needed title to their land and the construction of new irrigation works to irrigate a wheat crop. Both these objectives, together with the desire of landless colonists to acquire land, coincided in the petition signed by some eighty persons of Spanish origin living within the jurisdictions of Irapuato and Yuririapúndaro, to found the villa of Salamanca (Figure 3.1).[9]

The viceroy Gaspar Zuñiga y Acevedo, Conde de Monterrey, authorized the foundation of the villa in 1602. He explained that the petitioners

> reported to me that they lived scattered and isolated one from another in their haciendas and farms, without order nor local government, and that, although they sought for many years to join and congregate in some place in the region, it was not done because they had not found a place so suitable and adequate as that which they have discovered between the villas of Celaya and León, in the area called Valtierra, next to the estancia of Barahona and the great river that comes from Toluca, from which through experience and actual demonstration it is possible to channel water for irrigation of the lands on the river bank that are so dry as to be useless and of little value, and it is possible to irrigate not only these lands, but fifteen leagues of land, channelling the water by canals and dams that they offer to build at their own cost.[10]

The founders of Salamanca plainly envisaged it as a kind of collective irrigation project. But curiously, the proposed site did not lie in the Valle de Santiago, which could be irrigated with relative ease, but on the northern bank of the Lerma. The river here is joined by the River Laja as it traces a large "U" and curves to the west. Water was to be diverted on the east bank of the river some distance to the south of the bend and conducted by wooden aqueducts over the River Laja and a second large arroyo to fields above the bend of the river. Eventually the wood structures would be replaced by masonry. The precise course of the proposed aqueduct cannot now be charted, but, as Basilio Rojas has written, it was a project conceived in "Roman proportions."[11]

The selection of this site clearly had to do with the willingness of ranchers to give up their land to establish the villa. The owners of the estancias on the north bank were

---

[8]Gonzales, pp. 232-33. This very interesting book contains two pages on the Valle de Santiago clearly drawn from documentary sources that no longer exist.

[9]Bakewell, p 61. For a perceptive interpretation of the motives for the foundation of Salamanca, see Rojas (1976).

[10]Título y Fundación, p. 713.

[11]Rojas (1976), p. 70.

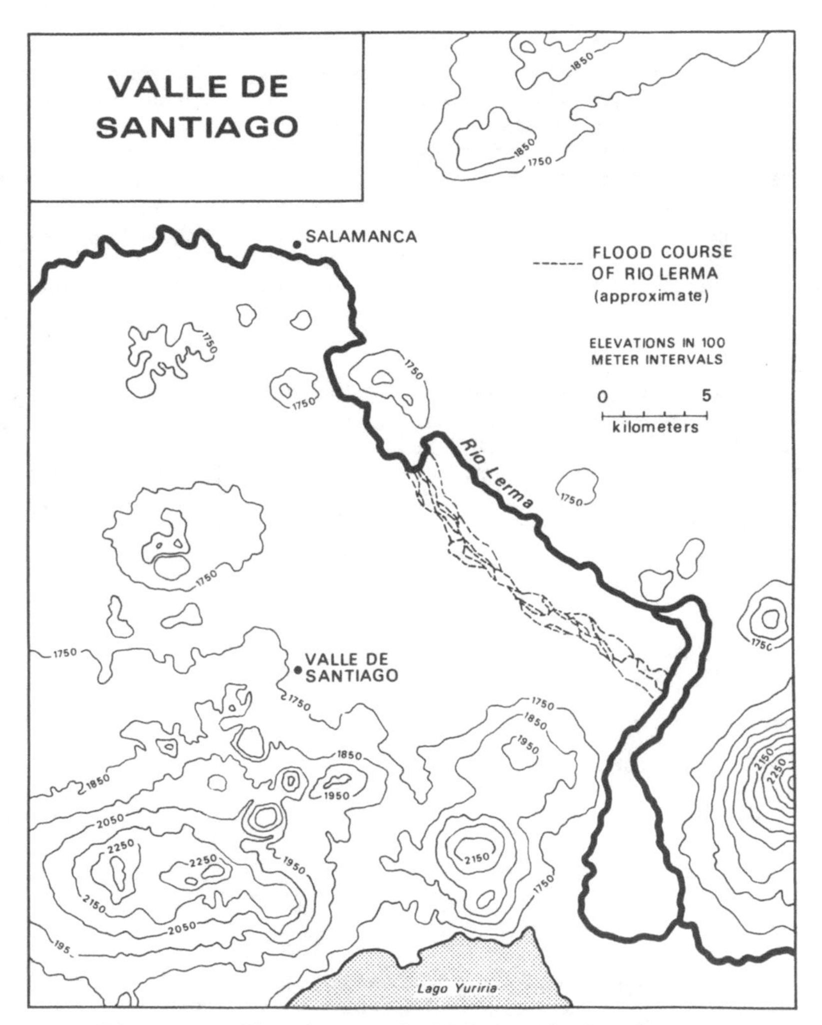

Figure 3.1 Sketch map of the Valle de Santiago

amenable, but only on the condition that they were compensated. The order of foundation provided for an arbitration procedure to determine the value of the land taken for vecindades, and it required that the vecinos offer adequate security for payment before taking possession. The order provided that the villa could grant thirty vecindades. Those granted to the original settlers would consist of four caballerías, four plots (suertes) for gardens, vineyard, and olive orchard and "the necessary water to be taken from the said river." The vecindades granted to vecinos admitted at a later date would be one-half this size. The vecinos would be required to live on their vecindades for ten years with no unauthorized absences of more than four months. The villa was authorized to channel water from the Lerma and to assess vecinos for the cost of building an irrigation system. Any vecinos failing to pay could be excluded from the villa and dispossessed of their vecindad.[12]

The founders of the villa secured special permission to form a cabildo immediately without having to wait a full four months until the beginning of the next year.[13] They prepared a plan whereby the vecinos' payment for their vecindades would be secured by a kind of purchase money mortgage. The plan was, however, rejected by the audiencia, which insisted that the vecinos offer security of payment before taking possession.[14] In face of this obstacle, the organization of the villa seems to have gone into temporary abeyance. But in 1606, the cabildo prepared a grandiose plan for distributing vecindades in sixteen districts within the jurisdiction of the villa. The Valle de Santiago was the fifteenth of these districts.[15] To enlist cooperation, the cabildo sought to enforce by judicial means the written commitments of a number of settlers--forty persons by one account, 150 by another--to join the villa and to assume a proportionate share of the costs of foundation.[16] But these efforts seem to have come to naught. There is a record of only two land grants on the north bank of the Lerma before 1616. Although the records of other grants may have been lost, much land immediately surrounding the villa still had not been distributed in 1619.[17]

The founding of the villa, however, served well the interests of the farmers in the Valle de Santiago who soon received a number of land grants. In 1607, the cabildo surveyed eleven recently issued grants, both full vecindades and smaller parcels, near the

---

[12]*Título y Fundación,* passim.

[13]AGN-G, Vol. 6, Exp. 275.

[14]AGN-G, Vol. 6, Exp. 312, 385, and 634.

[15]Gonzales, p. 232.

[16]The order of foundation mentions forty persons committed to join the villa. *Título y Fundación,* p. 715. A description of an attempt to enforce some 150 written commitments is found in AGN-T, Vol. 118, Exp. 1, f. 163v.

[17]Earliest recorded grants on northern bank: AGN-T, Vol. 118, Exp. 1, Cuad. 2, ff. 7, 209; AGN-M, Vol. 31, f. 319v. Proposal to create vecindades immediately adjacent to villa in 1619: AGN-M, Vol. 35, f. 21v. Land grants on the north bank in the period 1616-17: AGN-M, Vol. 31, ff. 100v, 190, 288v, and 319v; Vol. 32, ff. 236 and 237.

canal of the Brazo de Moreno.[18] The full vecindades included a right to ten days of water, which was later raised to twelve days.[19] There is no record that the villa encountered any difficulties in distributing this land, although a few persons were required to relinquish a claim to other land in exchange for a vecindad. Most likely, the grants merely served to legitimize the title of farmers who already occupied land in the area.[20]

The newly formed villa also provided a vehicle by which the farmers could improve the irrigation system. The cabildo employed a builder (*carpintero*) to construct a dam and a canal; and a landowner in the Valle de Santiago, Juan Fernández, was put in charge of the project.[21] The vecinos quite possibly had forced Indian labor at their disposal; in 1602, the Count of Monterrey had assigned Indians from Ucareo and Tajimaroa in Michoacán to assist in the building of houses and irrigation works in the villa of Salamanca. Their assignment to Salamanca was intended to be temporary, but we have it on the generally credible word of Pedro Gonzales, who read municipal documents destroyed in the Revolution, that the Indians were employed in the construction of the dam of the Valle de Santiago.[22]

The dam was completed by 1607. Two years later the cabildo addressed the problem of deepening and widening the canal. The vecinos in the valley were ordered to clear and widen by one-third the portion of the canal passing by their fields and, when this work was finished, to clear collectively the length of the canal lying between the valley and the dam.[23] Numerous references to the canal in subsequent land grants attest to its continued importance.[24]

Although one cannot trace the exact route of the canal, there is every reason to believe that, as in later times, the vecinos channeled water partly by canal and partly by a

---

[18]Gonzales, p. 233. There is an independent record of seven of the eleven persons mentioned by Gonzales. AGN-M, Vol. 26, ff. 139, 148v, 160v; Vol. 31, f. 248; Vol. 33, f. 16; AGN-T, Vol. 119, Exp. 1, Cuad. 2, f. 211; *El Obispado de Michoacán*, p. 87 (Luis Fanseco). At least two other land grants near the canal were made before 1607. AGN-T, Vol. 311, Exp. 2, f. 3; AGN-M, Vol. 26, f. 161; Vol. 31, f. 160v. For location of these two grants, see AGN-M, Vol. 31, f. 248.

[19]AGN-T, Vol. 2959, Exp. 141, f. 20v.

[20]AGN-M, Vol. 26, ff. 157v, 161.

[21]Gonzales, p. 233; AGN-M, Vol. 37, f. 27v; Appendix 1.

[22]AGN-G, Vol. 6, f. 215.

[23]Appendix 1.

[24]AGN-M, Vol. 30, ff. 100, 100v; Vol. 31, ff. 248, 248v, 260v; Vol. 33, f. 16; Vol. 35, f. 164; Vol. 37, ff. 41, 41v, 42.

flood course of the Lerma.[25] Within the Valle de Santiago the canal was described as following the four league jurisdictional limit of the villa of Salamanca.[26] The calculation of the jurisdiction appears mistaken, but the description at least suggests that the canal ran in an east-west direction. The excess water flowed through a drainage canal into the old river course that joined the Lerma at the northern end of the valley.[27]

Any identification of the location of the dam is hazardous, but it seems to have been a short distance south of the valley.[28] The earliest documents mention the farmers' dam near an "island" and the "diversion of water opposite the cerro de Uruetaro y Camémbaro."[29] This description fits the location of the estancia Uruetaro, later known as the Hacienda de la Bolsa, immediately to the south of the Valle de Santiago where the Lerma divides into several courses.[30] Pedro Gonzales refers to it as the dam of Santa Rita, but it was in a quite different location and served a different purpose than the later colonial dam of Santa Rita opposite Lake Yuriria.[31]

The cabildo continued to actively grant vecindades within the valley for about twenty-five years. We have a record of those grants that were later confirmed by the audiencia; twenty-six vecindades were approved by the audiencia before 1620 and ten were approved after that date.[32] Many of the new vecindades lay along the canal. When the alcalde ordinario of Salamanca surveyed the vecindades on December 17, 1609, the properties are described as all having the principal canal "at their head." Vecindades had 1,100 varas of frontage on the canal; half vecindades had 550 varas.[33] The properties

---

[25]Gonzales, p. 233. The principal flood course crossing the northeastern edge of the valley was known as the Brazo de Moreno. See AGN-M, Vol. 30, f. 99; Vol. 36, f. 35v; Vol. 70, f. 109; AGN-T, Vol. 311, Exp. 2, f. 30v; Vol. 586, Exp. 8, f. 89; Vol. 1107, Exp. 1, Cuad. 1, f. 95v, Cuad. 4, f. 8v; Vol. 1114, Exp. 1, f. 8. The name, however, was sometimes applied generally to this section of the valley. The lower reaches of the flood course were also known as Río Viejo or arroyo de Sotelo. See Figure 3.2.

[26]AGN-T, Vol. 294, Exp. 1, f. 92v; Vol. 311, Exp. 2, f. 3.

[27]AGN-M, Vol. 33, f. 78; AGN-T, Vol. 2959, Exp. 141, ff. 7v-8.

[28]AGN-T, Vol. 2959, Exp. 141, f. 7v; Appendix 1 (order to clean canal above valley).

[29]AGN-M, Vol. 37, f. 27v, and Appendix 1.

[30]AGN-T, Vol. 311, Exp. 2, f. 15, 30v; Vol. 432, Exp. 1, Cuad. 3, f. 40v (location of estancia Uruetaro). For location of head of canal in eighteenth century, see AGN-T, Vol. 2959, Exp. 141, f. 24v.

[31]Gonzales, p. 233. Gonzales tells us that the dam was built at a cost of 15,000 pesos and was financed by the sale of 150 shares (acciones) of 100 pesos--an account that also suggests confusion with a later dam. Figure 3.3 shows a dam at the head of the canal serving a different function from that of the dam of Santa Rita. For description of the dam of Santa Rita, see AGN-T, Vol. 114, Exp. 1, f. 28; AGN-C, Vol. 164, Exp. 4.

[32]The following land grants before 1620 can be identified as lying within the Valle de Santiago: AGN-M, Vol. 26, f. 119, 139, 145, 148v, 157v, 160v, 161; Vol. 28, f. 64; Vol. 30, f. 99, 100(2), 100v; Vol. 31, ff. 160v, 239v, 240, 248, 248v, 260v; Vol. 32, f. 166v, 167v, 208v; Vol. 33, ff. 10, 10v, 16(2), 78. In 1620 and after: AGN-M, Vol. 35, ff. 122, 164; Vol. 36, f. 35v(2); Vol. 37, ff. 41, 41v(2), 42, 166; Vol. 39, f. 222v.

[33]AGN-T, Vol. 2959, Exp. 141, ff. 7v-8.

thus were rectangular in shape with a length four times their width. A series of other vecindades were created along the Brazo de Moreno, largely beyond the reach of the central irrigation system.[34]

The large majority of the land grants were for less than a full vecindad, but the cabildo did not always adhere strictly to the scheme of vecindades and half vecindades contained in the order of foundation. In addition to the standard allotments, it granted vecindades of five or three caballerías and with eleven, eight, or four days of water. All the vecinos probably received water rights with their vecindad,[35] but the cabildo made no grants of water rights independent of grants of land. One cannot deny the generally egalitarian pattern of distribution, but certain family names, such as Guzmán and Fernández, recur in the grants, suggesting that there may have been more consolidation of landholdings than appears on the surface. For example, both Pedro Fernández del Rincón and his wife, María de Guzmán, received grants, totaling nine caballerías and twenty-one days of water.[36] A few vecinos were also privileged to receive grants of ranching land as well as agricultural land.[37]

The establishment of an agricultural community in the Valle de Santiago attracted persons residing north of the Lerma. In the mid-seventeenth century, the cabildo of Salamanca was heard to complain that Spaniards had forsaken the villa to live in the Valle de Santiago several leagues away.[38] The new haciendas in the valley similarly attracted Indians from Yuririapúndaro who sometimes later refused, with the complicity of the Spanish farmers, to respond to official summons to provide mandatory days of labor elsewhere. This phenomenon helped fuel jurisdictional disputes between Salamanca and the pueblo of Yuririapúndaro. Part of the new area of settlement lay beyond the original jurisdictional limit of Salamanca, but the villa nevertheless maintained its jurisdiction over the valley.[39]

The two earliest tithe reports, for the years 1631 and 1637, reveal a patchwork of small wheat farms. Most farms produced between 450 and 1,200 fanegas; only four produced significantly more. The bulk of the production was within the central irrigation system,[40] but five farms along the Brazo de Moreno produced ten to twenty percent of the

---

[34]AGN-M, Vol. 30, ff. 99, 100; Vol. 31, ff. 160v, 239v, 240; Vol. 33, ff. 10, 10v, 78; Vol. 36, f. 35v, Vol. 37, f. 166.

[35]The transcript of the grant to Juan de Cázeres does not mention water rights, but he appears elsewhere as the owner of six days of water. AGN-M, Vol. 30, f. 99; AGN-T, Vol. 586, Exp 8, f. 89v. Grants of land described as fronting on the canal probably imply a grant of water rights.

[36]AGN-M, Vol. 26, ff. 157v, 160v.

[37]AGN-T, Vol. 155, Exp. 4, f. 4v; AGN-M, Vol. 32, ff. 166v, 167v.

[38]AGN-T, Vol. 118, Exp. 1, ff. 6-6v.

[39]AGN-T, Vol. 294, Exp. 1, Cuad. 1. Folio 87 is of considerable interest for the history of labor in Mexico, because it very graphically supports the thesis that Indians migrated to haciendas in the seventeenth century to escape oppressive conditions in their pueblos.

[40]El Obispado de Michoacán, p. 87; ACM, leg. 860 (1637). Only six of the wheat farms can be linked by documentary references to the Valle de Santiago (AGN-T, Vol. 586, Exp. 8, f. 96; AGN-M, Vol. 26, f. 139; Vol. 31, f. 248v; Vol. 32, ff. 208v, 237; Vol. 37, f. 41) but, in view of the history of the northern bank of the Lerma, it is highly likely that the others were also isolated there.

crop. Aggregate production appears to have been in the order of 21-24,000 fanegas. In the decades after 1655, for which we have a nearly continuous series of tithe records, the valley seems to have experienced a rise in production that was followed by a slump in the last years of the century. In 1664, wheat production reached a peak of 45,000 fanegas. These early levels of production probably reflect the creation of branch canals. Much of the valley was served by long branch canals in the eighteenth century, but there is no evidence that they were built in that era. One of the more important such canals, Villadiego, clearly had its origin in 1612[41] and another, Las Rosas, can be placed in the middle of the seventeenth century.[42]

The villa of Salamanca itself scarcely existed in physical form in the early seventeenth century,[43] but around 1617 the plan to build an irrigation canal to the site was revived. The cabildo granted vecindades on the north bank with water rights[44] and contracted with one Martín de Veriga, master of architecture, to build a dam on the Lerma. It agreed to pay 16,000 pesos and give him six caballerías of land for the work, but, since the villa lacked the resources to keep its end of the bargain, it petitioned to the audiencia for permission to grant fifty vecindades, without obligation of personal residence in the villa. Each recipient of a vecindad would be required to pay "a sufficient sum to build the said intake and dam." Curiously, the need to compensate the ranchers occupying the land does not seem to have been an issue at this time. In 1619, the audiencia approved the issuance of thirty vecindades with the obligation only to build a house in the villa within two years.[45]

The plan seems to have met with little success. There is a record of only a handful of subsequent land grants on the north bank, and they disclose payment of only the nominal sums of 150 or 200 pesos "to aid in the construction of the dam and intake of water."[46] But somehow a dam was built. It probably had a wealthy backer in Luis de Cordoba Bocanegra, alcalde mayor of Celaya, whose family owned cattle estancias on the north bank.[47] In 1620, he secured a grant of two vecindades, twenty-four days of water, and a license to operate a flour mill, all of which plainly lay on the north bank. The grant was made in consideration of his aid in building the irrigation system on the river, and he received the right to take water from the head of the canal in preference to other

---

[41]AGN-T, Vol. 2959, Exp. 141, f. 18v.

[42]The canal Las Rosas can be identified with the desaguadero mentioned in the early grants. Moreover, the Hacienda Enmedio which existed in the mid-seventeenth century (AFC-CSC, leg. "titulos") was so named because it lay between the principal canal and the canal of Las Rosas. AGN-T, Vol. 432, Exp. 1, Cuad. 1, f. 127, Cuad. 2, f. 9v.

[43]AGN-T, Vol. 118, Exp. 1, f. 7; AGN-M, Vol. 35, f. 21v.

[44]AGN-M, Vol. 31, ff. 100v, 288v. The former can be placed with some assurance on the north bank (see AGN Mercedes, Vol. 35, f. 171). The location of the latter is somewhat conjectural.

[45]AGN-M, Vol. 35, f. 21v.

[46]The following land grants can be confidently located on the north bank: AGN-M, Vol. 35, ff. 134v, 171; Vol. 38, ff. 61, 63. The location of the following grants in this area is tentative: AGN-M, Vol. 35, f. 159v (Diego de la Maya and Juan Lopez).

[47]Alonso Pérez de Bocanegra is identified as the owner of an estancia on the north bank in the Order of Foundation. *Título y Fundación*, p. 714.

vecinos.[48] The dam appears to have been located within the jurisdiction of Salamanca, opposite the Valle de Santiago.[49] As suggested by Bocanegra's anxiety to secure priority in the use of the water, it was a failure from the beginning. In 1620 the canal was described as having "little water," and in 1644 Basalenque reported that the dam was "not used because of its defective construction."[50]

To avoid complete depopulation, the cabildo of Salamanca sought to enforce the conditions of vecindades requiring residence, or at least the construction of houses, in the town. In 1655, the cabildo also sponsored the foundation of an Indian pueblo, Santa María Nativitas, on the eastern edge of the villa and gave the Indians an expanse of land and the right to build a dam at their own expense "for the benefit of their land." But the Indian pueblo obviously could not succeed where the Spanish population had failed; it later obtained its water supply from wells.[51] Efforts to move the villa to the Valle de Santiago were resisted, and the town eventually acquired its own identity. Near the end of the colonial period, it was described by Humboldt as a "jolie petite ville, située dans une plaine."[52]

The north and east bank of the Lerma near Salamanca remained devoted to cattle estancias, interspersed with some maize cultivation throughout the colonial period.[53] The Rancho Potrero made use of the flood waters of the Laja to grow a wheat crop (trigo aventurero),[54] and the Hacienda Mendoza possessed an arroyo dam in the mid-eighteenth century that irrigated a small patch of land.[55] But irrigation was of marginal importance in the later colonial period. The tithe report of 1750 shows only six haciendas in Salamanca outside the Valle de Santiago that produced a small quantity of wheat in addition to maize.[56]

The Valle de Santiago itself lost some of its character as an irrigation community of small proprietors, but it retained a relatively fragmented pattern of land ownership and, despite threats of disorganization, a complex irrigation system. A remarkable document gives us a panorama of the valley in 1687-88.[57] The system of irrigation turns (tandas) based on grants of days of water was complemented by a system of irrigation boxes and gates (marcos) that regulated the flow of water into a farm during an irrigation turn. In

---

[48]AGN-M, Vol. 35, f. 73v. The text of the grant can be related in several ways to the north bank. Moreover, Bocanegra does not appear as an owner of land near the head of the canal in the Valle de Santiago. See AGN-T, Vol. 311, Exp. 2, f. 3.

[49]AGN-T, Vol. 118, Exp. 1, ff. 7v, 9, 157-62; El Obispado de Michoacán, p. 87.

[50]AGN-M, Vol. 35, f. 42; Basalenque, p. 314.

[51]AGN-T, Vol. 118, Exp. 1, ff. 6v-9v, 162v; Morin, p. 288.

[52]Humboldt, I, Book 3, Chap. 8, p. 286.

[53]AGN-T, Vol. 95, Exp. 6, f. 6v; Vol. 119, Cuad. 3, ff. 87-8, 144v; Vol. 879, Exp. 3; Vol. 1389, 2a pte, Exp. 1, f. 3v; Vol. 1447, Exp. 2, f. 2.

[54]MNA.AH, Colección General, leg. 3698, doc. 8.

[55]AGN-T, Vol. 275, Exp. 1, Cuad 3, ff 16-17.

[56]MNA.AM-ACM, roll 776636 (1750).

[57]The discussion of the Valle de Santiago in 1687-88 is taken from AGN-T, Vol. 2959, Exp. 141, ff. 10v-16v.

1687, water was distributed through thirty irrigation gates, but only four functioned properly, and farmers were known to open breaches (portillos) in the canals to take water without proper authorization. The assignment of irrigation turns itself was subject to dispute. Farmers on the upper portions of the canal were accused of taking water out of turn to the detriment of lower properties. Further problems were experienced in the maintenance of the system. The villa had enacted an ordinance that prescribed annual cooperative efforts to clean the system; the farmers were to provide peons to work on the canals for ten or twelve days each September. But through neglect of this obligation, the common portions of the canals--near the dam and drainage outlet--suffered siltation.

As owners of lower properties on the canal, the Augustinians secured an order of the audiencia that directed the alcalde ordinario of Celaya to intervene. He convened a general meeting of the farmers in January 1688 and demanded that the canals be cleaned and the gates be put in order within twenty days. In two subsequent meetings of the farmers in February, the alcalde confirmed that the work had indeed been completed, but to avoid further disputes, he prepared a comprehensive statement of the customary irrigation turns and ordered that they be respected. Beginning with the upper properties, the turns were described as follows:

> First tanda
>
> > The hacienda of the Lic. D. Nicolás Cavallero y Ocio that has 36 days in addition to the next property of which Nicolás Aguado is administrator that has 8 days, together make 44 days. This is one tanda and the first . . .
>
> Second tanda
>
> > The hacienda of D. Joseph Merino y Arevalo that has 8 days; that of D. Antonio de Medina Picazo that has 12 days, plus the right to one tercia of water to which he has priority; that of the Depositorio Gabriel de Valle Abarado, who has 20 days. These three add up to 40 days and comprise the second tanda . . .
>
> Third tanda
>
> > The hacienda of Regidor Nicolás Miguel in the first irrigation ditch has 12 days and in the second ditch has 6 days; Bernardino Guerra in the first and second ditches has 6 days of water; Diego Ramirez has 6 days of water; Nicolás de Rojas has another 6 days. Together these parties have 42 days and comprise the third tanda.
>
> Fourth tanda
>
> > Francisco Delgado Guerrero who has 6 days of water; the hacienda of the reverend mothers which Pedro García rents with 12 days of water; Juan García Muñiz for the farm of Pedro Garcia de León 6 days; the hacienda of Alferez D. Agustín Lejalde y Arisaga that has 20 days. This tanda has 44 days . . .
>
> Fifth tanda
>
> > The hacienda of the Tular that has 6 days; that of the reverend mothers of Valladolid that Philipe Vertiz occupies with 6 days of water; that of Francisco de Aguilar that belongs to the reverend mothers of Querétaro with 8 days of water; that of the Convent of Sr. San Agustín of the villa of Salamanca of which the administrator is the Rev. Fr. Fray Diego de Aguilar that has 24 days and together these parties and days of water add up to 44 days . . .

The use of the word "tanda" to denote a group of farms is unusual, but the meaning is clear: the water was distributed through five irrigation boxes placed at intervals along the canal. Each irrigation box received a turn of water, which was then distributed among the properties belonging to this subsystem. The document does not reveal the pattern of rotation among the irrigation boxes. Perhaps each farm received three irrigation turns a season to permit the customary three irrigations, but certain language suggests that the boxes received successively four days of water. In either case, the property owners belonging to each tanda faced the necessity of allocating their annual rights to days of water among the shorter periods when an irrigation box received water during a season. The turns commenced with upper and lower properties on alternate years. While the upper properties irrigated, the fifth tanda was given a small continuous supply (media naranja) for domestic use and for cattle. In the second tanda, the hacienda of Antonio de Medina Picazo received a continuous supply of water, one tercia, in addition to its irrigation turn, apparently as compensation for building a branch canal.[58]

In addition to these primary water rights, the alcalde ordinario confirmed complicated rules with respect to the excess water in the canal, described as the remaniente. Though the details are not entirely intelligible, the main features can be discerned. The first tanda did not receive remanientes for obvious reasons, but the fifth tanda also lacked the benefit of excess runoff, evidently because the drainage canal intercepted the water above the irrigation box. Both tandas were given an additional five days of water to compensate for their lack of remanientes. The second and third tandas possessed rights to fewer days of water than the fourth and fifth tandas. To adjust for this discrepancy, a system of irrigation turns was established for the distribution of remanientes themselves.

An irrigation superintendent (mayordomo de las acequias y presas), paid a salary of 300 pesos, was charged with the maintenance and operation of the system. Small properties contributed one peon for the annual cleaning of the canals; larger properties were assessed at a rate of one peon for every six days of water. The mayordomo was authorized to fine any landowner six pesos for refusing to send a peon when needed. A larger fine of 500 pesos was provided for general compliance with the order of the alcalde ordinario.

The document suggests a degree of social transformation and consolidation of land tenure. In place of the small farms of two generations earlier, one finds haciendas rented by absentee landlords. The tithe record of 1687 lists thirty wheat farmers[59]--precisely the same number as the number of irrigation gates mentioned in the document--but only five of these farmers appear as owners of properties in the judicial allocation. A large hacienda, owned by Nicolás Cavallero y Ocio, had appeared at the end of the canal, and there were five church estates (including a Carmelite hacienda on the Brazo de Moreno). The judicial description of the five tandas may not be an entirely comprehensive list of irrigated properties; the list of farms obliged to contribute peons for the maintenance of the canal includes additional references to the municipal land, three haciendas in the Brazo del Moreno, and a number of small ranchos; one of the ranchos is described as having a right to three days of water not included in the body of the order. But the total number of landowners clearly had diminished substantially.[60]

During the eighteenth century, the Valle de Santiago continued to be a region of small wheat haciendas. One hundred and fifty years after the original distribution of land, the characteristic size of an hacienda was five or six caballerías--not much larger than the

[58]See also AGN-T, Vol. 111, 2a pte, Exp. 2, f. 37; Gonzales, p. 233; Appendix 1.

[59]ACM, leg. 860.

[60]Thirty-six grants of vecindades were confirmed by the audiencia of which number at least thirty-one and probably all included water rights. See footnotes 32 and 35. Many other grants were never submitted to the audiencia for confirmation. A record of three such grants appears in later litigation. AGN-T, Vol. 118, Exp. 1, Cuad. 2, f. 221; Vol. 586, Exp. 8, f. 96; Vol. 2959, Exp. 141, f. 9. Other vecindades are encountered only in boundary descriptions, e.g., AGN-M, Vol. 26, f. 145 (Francisco Lucas).

original vecindades.[61] The largest hacienda, San Rafael, was only seventeen and one-half caballerías.[62] In the mid-nineteenth century, the valley was said to contain twenty-eight wheat haciendas,[63] and it retained the field pattern of narrow rectangles, formed by the shape of the original grants at the beginning of the seventeenth century.[64] This division of land ownership still permitted the existence of numerous small rented ranchos; tithes (including a part of the wheat crop) were collected in 1800 from some forty to fifty ranchos located mostly on three haciendas.[65] But the apparent persistence of small holdings is somewhat of an illusion. While many small haciendas retained a separate identity, they were frequently connected by common ownership with other properties, often contiguous properties, and many were held by ecclesiastical organizations. For example, four neighboring haciendas came under the control of the Jesuits; although each hacienda was quite small, together the properties comprised fourteen caballerías.[66] Absentee ownership was the general rule; only five haciendas in 1793 were administered by their owners. Links between the absentee landowners and the cabildo of Salamanca could not have been strong, but there is record of three landowners in the late eighteenth century who were in fact municipal officers of Salamanca.[67]

In 1707, the Augustinians purchased from Juan Caballero Navarro sixty-seven caballerías of land, thirty-six days of water, two cattle estancias, and eleven criaderos, all lying to the south of the valley. The land included the Hacienda San Rafael at the head of the canal and the former Estancia of Uruetaro, then known as the Hacienda de la Bolsa, lying just beyond the irrigation system of the valley, which Caballero had begun to

[61]Santa Ana, 5 caballerías in 1771 (ACM, leg. 837, Exp. 497); Pitayo, 3 caballerías in 1734 and 6 caballerías in 1762 (ACM, leg. 823, Exp. 359); San José de Jaral, 6-1/2 caballerías, and Mesquite Grande, 5-1/2 caballerías in 1759 (ACM, leg. 831, Exp. 450); Guadalupe, 6 caballerías, Hoya Blanca, 5-1/2 caballerías, and San José, 9 caballerías in 1752 (ACM, leg. 830, Exp. 425); Guadalupe, 3 caballerías, and Baullos (?), 6 caballerías in 1762 (ACM, leg. 837, Exp. 495); Hacienda de Carmelitas, 8 caballerías in 1712 (LADC, WSB 463); San José de Sintano, 3 caballerías in 1705 (MNA.AM-Qro, roll 27, Registro Público de la Propiedad, Lib. no. 1, item 140).

[62]ACM, leg. 837, Exp. 495.

[63]AGN-A, Vol. 97, Exp. 2, f. 28v.

[64]Romero, p. 67.

[65]Morin (1979), pp. 272-76.

[66]For Jesuit haciendas, see ACM, leg. 833, Exp. 432; AGN-T, Vol. 432, Exp. 1, ff. 6v-13v, 80v, 129. For other examples of consolidation or common ownership, see ACM, leg. 823, Exp. 359 (increase in size of Hacienda Pitayo); ACM, leg. 837, Exp. 497 (common ownership of Hacienda Santa Ana and Hacienda Grande); ACM, leg. 862 (two groups of haciendas with common owners in 1758: Villadiego, Tular, and Huérfanos; San José Sintano, Los Lobos, El Cerrito de Cal, and Cerro Prieto); AGN-T, Vol. 1107, Exp. 1, Cuad. 1, f. 47 (map: Casas Blancas and Paramo with common ownership in 1784); ibid, Cuad. 4, f. 8, and private archive of Basilio Rojas (Terán and Pitayo with same owners in 1804).

[67]Morin, p. 281; Rojas (1971), p. 8. Owners who were regidores of Salamanca: AGN-T, Vol. 1107, Exp. 1, f. 47 (Tomas Machuca); Vol. 1113, Exp. 1, Cuad. 6, f. 10 (Simón Tabera); LADC, WBS-463, f. 1v (Martínez Conejo).

convert into an irrigated agricultural estate.[68] The transaction would have a vital influence on the later history of the Valle de Santiago. As owners of the Hacienda San Rafael and several smaller haciendas, the Augustinians were now the largest landowners in the valley and, more important, the owners of much of the rich agricultural land along the west bank of the Lerma between the Valle de Santiago and Lake Yuriria.[69]

The water for irrigation continued to be distributed through central irrigation boxes.[70] The valley was no longer divided into five tandas, but one finds various other divisions. For example, an agreement in 1780 describes the valley as comprising four irrigated sectors. Three irrigation boxes--Terán, Villadiego, and Las Rosas (or Santa Rosa)--were of central importance in regulating the flow of water through long branch canals (bearing the same names) to haciendas to the north.[71] The canal of Villadiego alone served eight or ten haciendas.[72] A few properties lying between these subsystems received water from more than one branch canal.[73] The number of water users seems to have remained rather constant. About twenty haciendas usually produced the bulk of the wheat crop but the tithe reports mention other very small producers; most were undoubtedly tenant farmers.[74]

The level of wheat production in the eighteenth century was generally comparable to that of the previous century. It rose above 60,000 fanegas in 1725 but usually tended to fluctuate around 30,000 fanegas.[75] Maize cultivation had been minor in the early period; in 1664, the valley produced only 4,160 fanegas and in 1665, only 5,030 fanegas. But in the eighteenth century, maize production, presumably stimulated by the demand of the Guanajuato mines, rose steadily and dramatically. By the end of the century, the annual harvest reached 60-80,000 fanegas. Was this increase in production assisted by irrigation? Only one colonial document mentions the irrigation of maize; it indicates that

---

[68]AGN-T, Vol. 311, Exp. 2, f. 2v; Vol. 432, Exp. 1, Cuad. 1, f. 130v, Cuad. 4, ff. 39-40v, Cuad. 5, f. 2.

[69]Other Augustinian haciendas in valley: San Miguel in 1755 (AGN-T, Vol. 2959, Exp. 141, ff. 4, 6v); Guadalupe and Baullosa in 1762 (ACM, leg. 832, Exp. 495); Guantes, San Xavier, and San Antonio in 1758 (ACM, leg. 862).

[70]Descriptions of the principal canal: AGN-T, Vol. 394, Exp. 2, f. 57; Vol. 432, Exp. 1, Cuad. 1, f. 7, Cuad. 2, f. 9; Vol. 2959, Exp. 141, f. 24v.

[71]Canal Villadiego: AGN-T, Vol. 394, Exp. 2, f. 57; Vol. 2959, Exp. 141, ff. 1-4, 24v. Canal las Rosas (or Santa Rosa): AGN-T, Vol. 432, Exp. 1, Cuad. 1, ff. 29v, 118, Cuad 2, f. 9; Vol. 2959, Exp. 141, ff. 6v, 24v. Canal of Terán: AGN-T, Vol. 586, Exp. 8, f. 44; Vol. 1107, Exp. 1, f. 95v. Division of valley in four irrigated sectors: AGN-A, Vol. 97, Exp. 2, f. 10. In two sections: AGN-T, Vol. 1107, Exp. 1, f. 131.

[72]AGN-T, Vol. 2959, Exp. 141, f. 24v.

[73]E.g., AGN-T, Vol. 586, Exp. 8, f. 44; Vol. 1107, Exp. 1, f. 45v.

[74]MNA.AM-ACM, roll 776636 and 776638 (years of 1750 and 1800); ACM, leg. 862; AGN-M, Vol. 70, f. 109 (more than eighteen farms in 1722). In 1861, there were twenty-eight irrigated haciendas in the valley--a number that corresponds remarkably to the thirty irrigation gates mentioned at the close of the seventeenth century, but is subject to various possible explanations. AGN-A, Vol. 97, Exp. 2, f. 28v.

[75]ACM, leg. 860.

in 1772 there were irrigation turns in the spring to benefit the newly sown maize crop.[76] Most of the increase in maize production was probably the result of the conversion of rich grazing land to tillage, but we may surmise that irrigation also played a part in raising production.

In 1721, the valley underwent another effort to confirm and clarify water rights. The landowners petitioned the audiencia to commission a judge to settle difficulties. Their petition was reviewed by the scholar, Sáenz de Escobar, whom we will meet again in the chapter on technology.[77] Sáenz recommended the construction of irrigation boxes in a manner that would reduce variations in flow resulting from differences in the head of water above the gates. There is no way of knowing whether Sáenz' ideas were carried out, but a judge was commissioned, and, with his approval, the landowners met "to assign to each farmer the precise days and irrigation turns in which each one is to receive water from the river through the irrigation boxes." They succeeded in working out a written agreement, later ratified by the viceroy, that all water users were induced to sign. Among other matters, the agreement precisely fixed the size of irrigation gates in terms of surcos. A later agreement in 1780 seems to have varied the standard; the owner of three caballerías of land was given the right to a gate one-third of a vara square.[78]

Although the colonial records of Salamanca have been lost, we know that ordinance "five" prescribed annual meetings of landowners to consider irrigation problems.[79] There is no doubting the importance of these meetings.[80] In the late eighteenth century, the proceedings were actually recorded by a scribe in a sort of minute book. We have a transcription of the minutes of one meeting held on April 3, 1784, near the end of the growing season when some haciendas suffered from an inadequate supply of water. Representatives of the landowners met in the house of the alcalde ordinario of Salamanca. They affirmed the rights of certain parties to assigned shares of water and granted short irrigation turns as a special dispensation to a few haciendas experiencing exceptional need.[81]

Other matters considered in the meetings of landowners included penalties for unauthorized breaches in the canal, the destructive use of canals as watering places for cattle, and, above all, the clearing of canals. The haciendas contributed pro rata according to the number of caballerías of irrigated land they possessed (not days of water as in the seventeenth century); the work always proceeded from the head of the canal downward ("cogiendo de arriba a bajo"). Haciendas served by a branch canal were given joint responsibility for cleaning it--a task that periodically called for moderate expenditures. The Hacienda Santa Rosa complained that "it was obliged every three or

[76]AGN-C, Vol. 164, Exp. 4.

[77]The report of the fiscal is merely signed "Sáenz," but the context makes identification with Sáenz de Escobar virtually certain.

[78]AGN-M, Vol. 70, ff. 109-112; AGN-A, Vol. 97, Exp. 2, f. 10; Appendix 1.

[79]AGN-M, Vol. 70, f. 111.

[80]AGN-T, Vol. 394, Exp. 2, f. 57v; Vol. 2959, Exp. 141, f. 23v; AGN-A, Vol. 97, Exp. 2, ff. 10, 29; ACM, leg 830, Exp. 426, ff. 119, 134.

[81]AGN-T, Vol. 1113, Exp. 1, Cuad. 3, ff. 42-44v.

four years to spend three hundred pesos, more or less, to clear the canal called Río Viejo whose maintenance is assigned to the haciendas of this area."[82]

The administration of the irrigation system was supervised by a water judge (juez de aguas) who was no doubt the same municipal officer as the mayordomo de las acequias y presas of the seventeenth century. A document written in 1778 explains:

> In order to carry out its intent, the cabildo was accustomed to appoint a salaried vecino, who with the title of water judge looked after the maintenance and repair of the canals and sought to avoid waste, loss and theft of the water, punishing and placing in jail offenders.[83]

Although he was appointed by the cabildo, the salary of the water judge was paid by the landowners themselves. He actively removed unauthorized irrigation gates and ditch dams and destroyed the earthen dams on the Lerma that upstream farmers used to divert water before it reached the valley. But he was subordinate to the alcalde ordinario de Salamanca and sometimes appears as a kind of deputy of other municipal officers. He lacked any clear rule-making authority.[84] Conflicts that were not resolved in the annual meetings soon reached the alcalde ordinario of Salamanca.[85]

The basis for cooperation among landowners rested not only on ordinances and judicial orders by increasingly on agreements formalized in annual meetings or in written contracts. In addition to the general agreement of 1721, we will soon encounter important agreements in the years 1780 and 1797. The landowners could attempt to enforce these agreements by denying water to a particular farm (the annual payment of a fee for the water judge's salary was a condition to receiving water), but their decision did not carry any legal authority. Differences of interpretation or simple delinquency of an individual landowner often could be resolved only through judicial remedies. A particularly awkward situation could arise when a hacienda changed hands. The proper administration of the system required that the new landowner adhere to agreements of which he was not a party. By all accounts the administration of the system generated an inordinate amount of litigation in the nineteenth century--a problem that was clearly inherited from colonial times.[86]

The lower reaches of the branch canals, known as the "padron de abajo," came into contact with a more or less autonomous source of irrigation, the flood course known as the Brazo de Moreno, Río Viejo, or, in its lower reaches, the Arroyo de Sotelo. The arroyo flooded during the summer, and the flow could be easily augmented by ditches connecting it to simple dams on the Lerma. As we have seen, it was probably the scene of the first efforts of irrigation in the Valle de Santiago, and it was exploited throughout

[82]AGN-T, Vol. 394, Exp. 2, f. 58; Vol. 586, Exp. 8, f. 10; AGN-A, Vol. 97, Exp. 2, f. 10 et seq; ACM, leg. 830, Exp. 426, ff. 119, 134.

[83]AGN-C, Vol. 164, Exp. 4.

[84]AGN-T, Vol. 394, Exp. 2, f. 51; Vol. 1107, Exp. 1, Cuad. 1, ff. 2-3, 48, Cuad. 2, f. 21v; Vol. 1113, Exp. 1, Cuad. 1, f. 44; Vol. 2959, Exp. 141, ff. 2v-6v, 24v; ACM, leg. 830, Exp. 426, f. 119.

[85]E.g., AGN-T, Vol. 1427, Exp. 3, f. 1; AGN-A, Vol. 97, Exp. 2, f. 10.

[86]AGN-A, Vol. 97, Exp. 2, ff. 29v, 31; ACM, leg. 830, Exp. 426, f. 119; Romero, p. 67; Gonzales, p. 233.

the seventeenth century.[87] Litigation in the eighteenth century has given us an abundance of information about the lower portions of the arroyo in the northeast corner of the valley, including a meticulous drawing of the area.

Where the arroyo joins the Lerma on the rancho Pitayo, a large flour mill with six mill stones competed with upstream properties for water. Although the mill received some water from the canal of Terán, it depended on the flow in the arroyo to be in full operation. A small dam on the Lerma above the mill augmented the volume of flow by channeling river water directly to the arroyo. This dam may have been built in the early eighteenth century by the owner of the Hacienda Sotelo, which lay beyond the reach of the central irrigation system, but the historical record is unclear.[88] In the 1780s, the owner of the mill of Pitayo, in a bid to obtain exclusive use of the lower reaches of the arroyo, built new canals and ditch dams and closed off ditches leading to the Haciendas Casas Blancas, San Bernardo de Majadas, and Terán that had long irrigated both from the arroyo and the central irrigation system[89] (Figure 3.2). The incident seems to have been a repetition of many similar conflicts. One finds a history of litigation over the arroyo water between the Haciendas Sotelo and Las Majadas and between other parties whose dams cannot now be located.[90] The utilization of the arroyo clearly extended farther to the south. The Haciendas Huérfanos and Charcas, among others, irrigated from it, and in 1784, a dam was built on the Lerma near the Hacienda Cerrito del Puerco to divert more water to the arroyo.[91]

The easy access to water and the impermanent nature of the dams militated against any formal regulation of water rights on the arroyo. The flour mill of Pitayo probably could trace its rights to an early grant, but there were no other formal grants of water rights. The periodic meetings of landowners sometimes sought to regulate irrigation along the arroyo,[92] but waterworks were built at individual initiative and maintained at private expense.[93] Since a dam or canal that benefited one farm could flood property upstream or reduce the flow to property downstream, it is not surprising that the arroyo was the scene of interminable conflicts and constantly changing patterns of irrigation. While one dam might be forcibly destroyed, another would be built.

---

[87]The separate listing of properties in the Brazo de Moreno in the tithe report of 1631 suggests a largely autonomous status based on a separate water supply, and a document dated 1662 attests to the existence of dams on the arroyo. AGN-T, Vol. 2963, Exp. 49. The general location of the dams can be inferred from the document, but see also AGN-T, Vol. 586, Exp. 8, f. 96. For the Hacienda Sotelo, see AGN-T, Vol. 115, 2d parte, f. 107 (1683).

[88]AGN-T, Vol. 1113, Exp. 1, Cuad. 6, f. 2v; Vol. 586, Exp. 8, f. 46.

[89]For the voluminous history of this conflict and its antecedents, see AGN-T, Vol. 586, Exp. 8; Vol. 1107, Exp. 1, e.g., cuad. 1, ff. 81, 147; Vol. 1113, Exp. 1; Vol. 1114, Exp. 1.

[90]Sotelo and Las Majadas: AGN-T, Vol. 621, Exp. 4, and AGN-M, Vol. 75, f. 170v. Other dams: AGN-M, Vol. 81, f. 50v; AGN-T, Vol. 2963, Exp. 49.

[91]AGN-T, Vol. 1107, Exp. 1, f. 131; Vol. 1113, Exp. 1, Cuad. 3, f. 43v, Cuad. 6, ff. 39, 91, 93.

[92]AGN-T, Vol. 586, Exp. 8, ff. 96-7; Vol. 1113, Exp. 1, Cuad. 3, f. 43v; Vol. 1114, Exp. 1, f. 8.

[93]For maintenance of the dams supplying the mill of Pitayo, see AGN-T, Vol. 1107, Exp. 1, f. 46; Vol. 1113, Exp. 1, Cuad 6, f. 12.

Though generally well endowed with water, the Valle de Santiago suffered occasional shortages. The drought of 1780 was evidently severe enough to stimulate cooperation between the landowners in the valley and the Augustinians to develop a new source of supply.[94] After acquiring the Hacienda de la Bolsa north of the valley, the Augustinians had set about to convert it into a wheat hacienda. According to the Augustinian historian Fr. Nicolás Navarrete--unfortunately not a reliable source--around 1720 the order built a canal eighteen kilometers in length extending from "Santiago Maravatio" (sic).[95] Be that as it may, the fact that the Augustinians spent at least 11,000 pesos for a flour mill on the hacienda suggests that a canal of some kind was built from the south.[96] The stage was set for a further expansion of the system that would benefit both the Hacienda de la Bolsa and the Valle de Santiago: the exploitation of the water of Lake Yuriria.

The level of Lake Yuriria, a broad and shallow basin connected to the Lerma by a short canal, rose and fell with the water level of the river. To use the lake as a reservoir, it was necessary only to place headgates at the entrance to the canal, which could be closed during high waters and opened gradually as needed. The Lerma itself could be used to conduct the water most of the way to the Valle de Santiago, but, because the river divides into several channels, it was best to dig a second discharge canal north of the lake to a channel of the Lerma that led directly and exclusively to the valley. This discharge canal could be built in an existing water course formed by overflow from the lake during periods of exceptionally high water. Additional headgates, of course, had to be built on this canal. The existing dam on the channel of the Lerma near the Valle de Santiago could be used to divert water to the irrigation system (Figure 3.3).

In 1780, the landowners in the valley hired an expert to study this relatively simple project and met three times to consider the matter and other problems of the valley. The product of their deliberations was a fourteen-point written agreement, executed by all the landowners in the presence of the royal scribe of Celaya. The agreement put to rest a lawsuit between the landowners and the Augustinians, dealt in very general terms with several administrative matters, and established the terms for investment in the new canal. The Augustinians, regarded as the owners of only the eastern one-half of Lake Yuriria, were enjoined to secure the necessary rights to the remainder of the lake from the pueblo of Yuririapúndaro. Presumably in recognition of their title to the lake, they were given the right to a continuous supply of one tercia of water for the irrigation of the Hacienda de la Bolsa. The cost of digging the canal and building the headgates, estimated at only 5,000 pesos, would be borne by the landowners of the Valley de Santiago who would be assessed pro rata according to the number of caballerías they possessed.

Two of the administrative provisions of the agreement are of particular interest.

> Part 8. That the allocation...(of the water) must be in proportion to the caballerías that are comprised in the sections Rio Abajo, Marco de Terán, Santa Rosa and that of the upper properties.

---

[94]Florescano (1980), p. 23.

[95]Nicolás Navarrete, I, p. 454.

[96]AA, Lib. Prov., Vol. 5, f. 153.

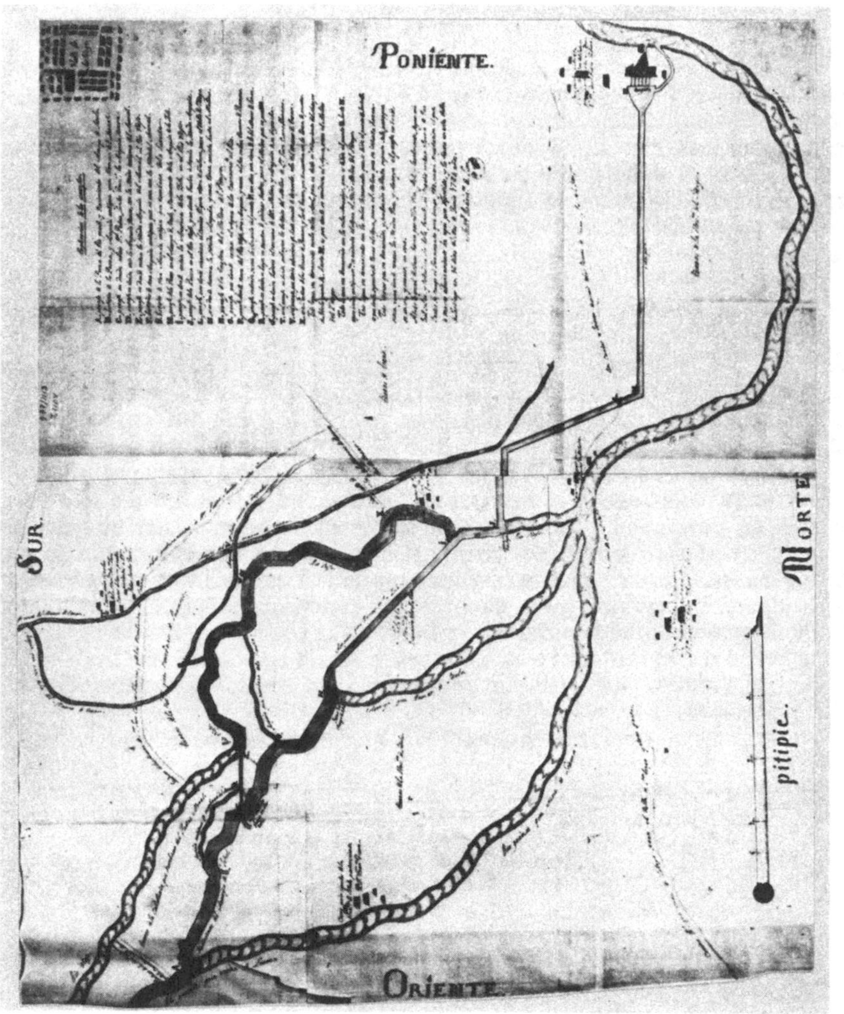

Figure 3.2 Northeastern fringe of Valle de Santiago
(Source: AGN-T, Vol. 933)

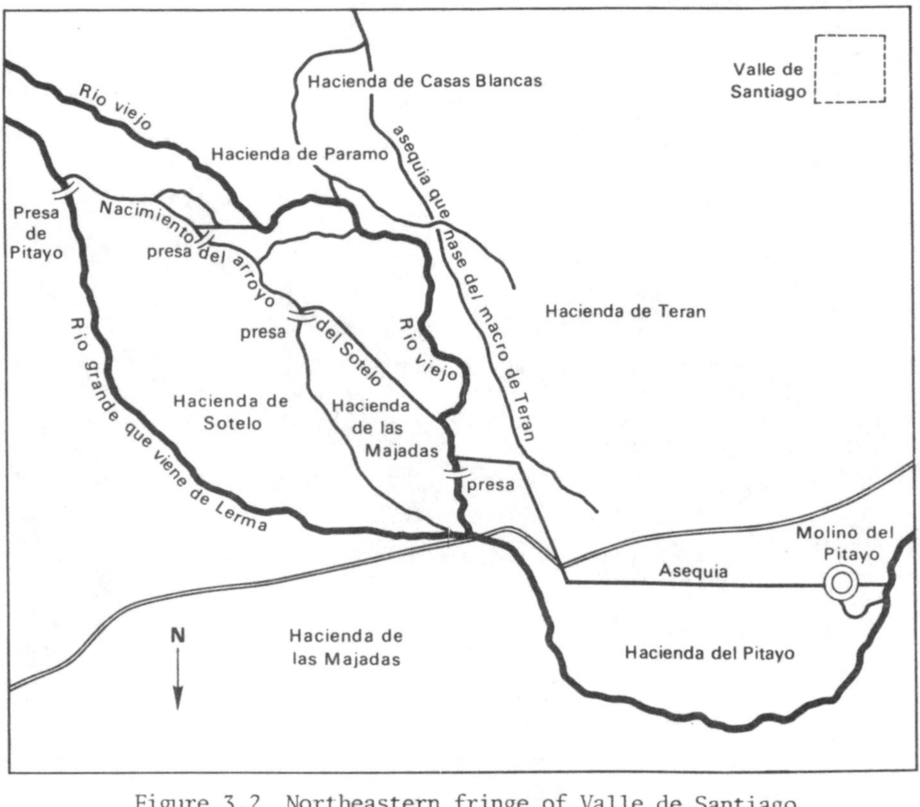

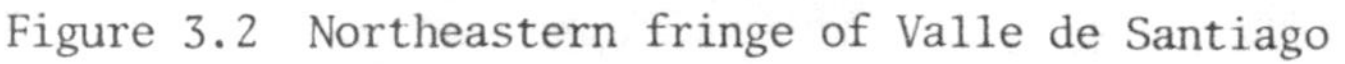

Figure 3.2 Northeastern fringe of Valle de Santiago

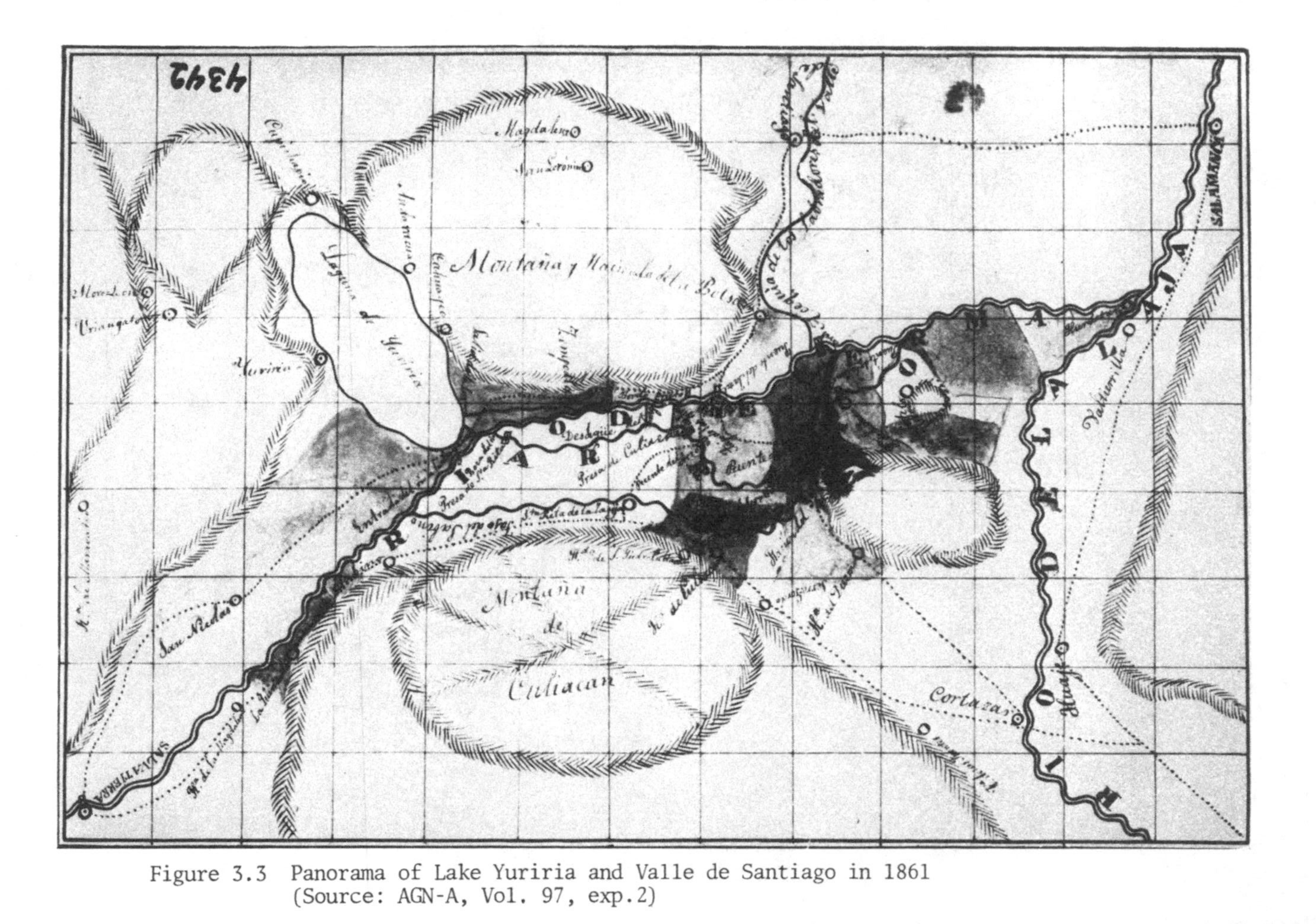

Figure 3.3 Panorama of Lake Yuriria and Valle de Santiago in 1861
(Source: AGN-A, Vol. 97, exp.2)

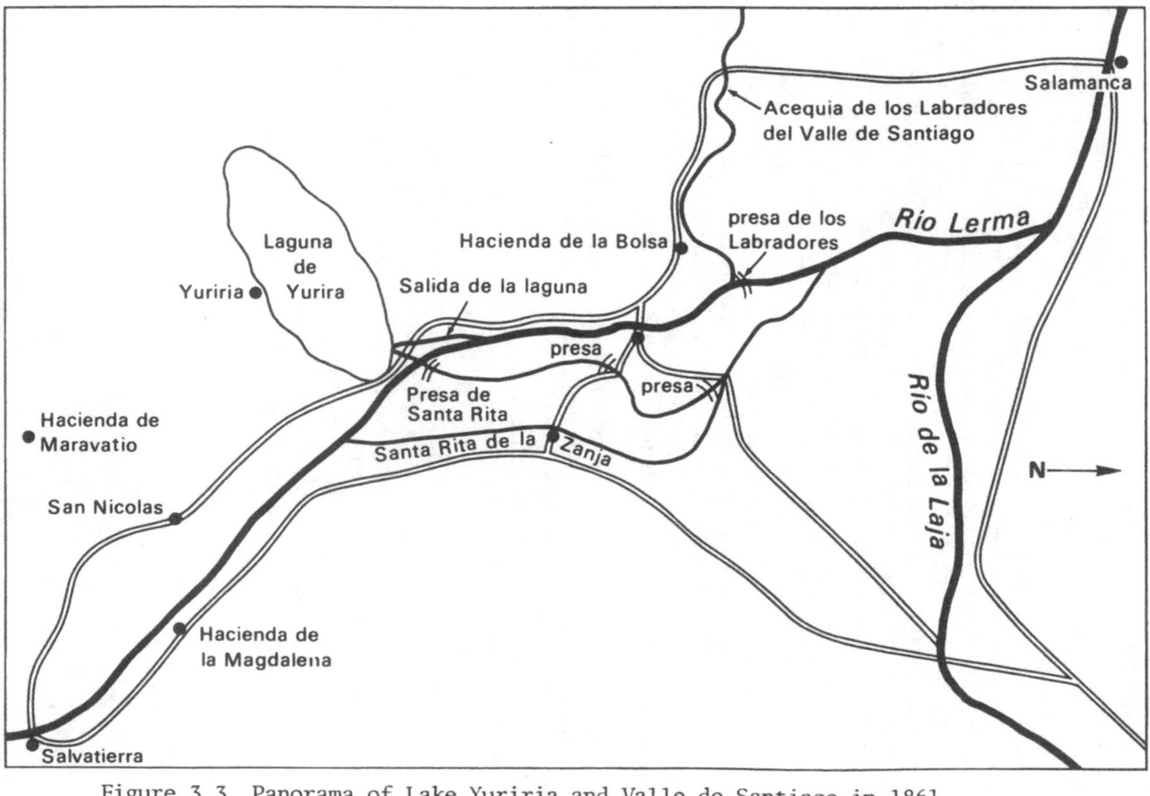

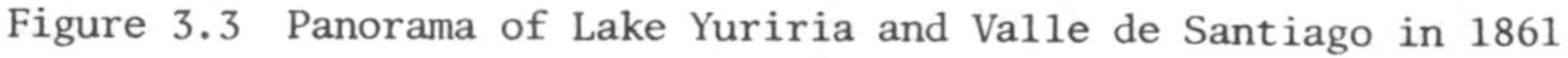
Figure 3.3 Panorama of Lake Yuriria and Valle de Santiago in 1861

> Part 9. That the irrigation turns must be employed successively from the 20th of February until the corresponding day of May...[97]

Clause eight suggests that the landowners had moved away from the allocation of water based on rights to days of water contained in the original land grants to a more practical system of allocation based on the amount of irrigated land each possessed. The ninth clause contains the information that in the late eighteenth century the irrigation turns began on February 20. For irrigation before that date, the irrigation boxes apparently effected a proportionate division of a continuous flow of water.

The farmers' new reliance on Lake Yuriria required them to consider the connection between that lake and Lake Cuitzeo. In exceptionally rainy seasons, Lake Cuitzeo overflowed into an old outlet leading to Lake Yuriria. The brackish waters of Lake Cuitzeo were sometimes fatal to the fish of Lake Yuriria and clearly unsuitable for irrigation. The landowners agreed to build a dike in the outlet from Lake Cuitzeo, but these harmful overflows continued to occur from time to time in the nineteenth century.[98]

Unfortunately, the storage of a useful amount of water in Lake Yuriria resulted in the flooding of an important part of the Hacienda San Nicolás for almost one-half the year. Although their properties in the Valle de Santiago were served by irrigation from the lake, the Augustinians sought to limit the extent of the flooding. The ensuing negotiations with the landowners of the Valle de Santiago resulted in an agreement in 1797. The landowners agreed to build a dike along the eastern side of the lake that would conform to precisely stated specifications; and upon construction of the dike, the Augustinians agreed to allow the lake to be used for the storage of water during specified times of the year. A stone was placed in the lake to measure the maximum level of water permitted. The Augustinians were given the right to close the canal headgates upon the landowners' default in the agreement.[99] According to Basilio Rojas, the dike took many years to build and marked the real beginning of the effective use of the lake as a reservoir for irrigation.[100]

The exploitation of Lake Yuriria as a reservoir curiously did not result in any expansion of irrigation in the Valle de Santiago, at least if we can take wheat production as a guide. The project was evidently a defensive measure to counteract declining water supplies. It is difficult to identify new upstream diversions of the River Lerma that might have substantially diminished the flow to the valley, but perhaps the western channel of the Lerma had silted up somewhat, increasing the flow to the eastern branches of the river that were not tapped by the Valle de Santiago. A phrase in the agreement of 1780 strongly suggests this explanation.[101] The landowners of the Valle de Santiago, nevertheless, felt threatened by upstream diversions of water and, especially, by the dam of the Hacienda Santa Rita near Salvatierra.

Portions of the rich land to the east and south of the valley, long devoted to stockraising, began to be put under irrigation in the eighteenth century. A survey of the haciendas in this area in 1793 notes in passing that none of the dams of "the

[97]AGN-A, Vol. 97, Exp. 2, f. 10 et seq.

[98]Ibid, ff. 10 et seq, 17, 66.

[99]Ibid, f. 51; Appendix 1.

[100]Rojas (1976), p. 75.

[101]AGN-A, Vol. 97, Exp. 2, f. 10 et seq (fatal estado del Rio Grande, por cuya mala disposicion, y haberse ensolvado sunamente, habien tomado distincta rombo las aguas).

aforementioned haciendas" were closed--an apparent reference to the use of floodgates--but it does not specify which haciendas actually possessed irrigation systems.[102] Other documents provide evidence of limited and irregular irrigation on the land east of the Valle de Santiago[103] and of an increasingly well-established irrigation system of the Hacienda Santa Rita farther to the south. In 1755, an appraiser commented that this hacienda's great potential for irrigation was only partially realized.[104] Later attempts to maintain and improve its irrigation system drew the hostility of the vecinos of the Valle de Santiago. The water judge of Salamanca tore down the hacienda's dams on many occasions ("muchas veces"), and the hacienda responded by destroying the wood and earth dams that the vecinos built to divert water to the western channel of the Lerma. In 1772, the vecinos brought a legal action against the hacienda, protesting its clandestine usurpation of water.[105] Despite this opposition, the Hacienda Santa Rita possessed a dam in 1793 that was permanent enough to be equipped with floodgates.[106] In the nineteenth century, the owners of the hacienda and the landowners of the Valle de Santiago would begin to work together for their mutual benefit by joining in the operation and maintenance of a dam that would not only better irrigate the rich land of the hacienda but also channel some of the flow of the east branch of the Lerma to the Valle de Santiago.[107] During the colonial period, however, the dam of the Hacienda Santa Rita was not yet integrated into the irrigation system of the valley.

By the beginning of the nineteenth century, the Valle de Santiago had evolved from an irrigation community of small farmers to a collection of small but productive properties at the edge of the great Augustinian domain to the south. The forms of cooperation among irrigators changed subtly. Formal written agreements assumed a key function. But the landowners remained well enough organized to maintain the early irrigation system and to invest jointly in a new source of supply in Lake Yuriria. Further joint investment with the Hacienda Santa Rita would soon follow.

---

[102]AGN-T, Vol. 1114, Exp. 1, f. 28v.

[103]See AGN-T, Vol. 602, 1a pte, f. 1 (suerte of wheat in Culiacán, 1734); ACM, leg. 833, Exp. 432, f. 39 (an evaluation that precludes extensive irrigation on Hacienda Savino, bordering Culiacán, 1753); ACM, leg. 862 (no wheat production in Hacienda Jaral, 1758); ACM, leg. 841 (small wheat production in most years, 1778-86).

[104]AGN-T, Vol. 787, Exp. 1, f. 300.

[105]AGN-C, Vol. 164, Exp. 4.

[106]AGN-T, Vol. 1114, Exp. 1, f. 28.

[107]Gonzales, p. 233; AGN-A, Vol. 97, Exp. 2, f. 10 et seq; Appendix 1.

4
Querétaro

The primitive origins of Querétaro at the eastern edge of the Bajío can be found in the first decades after the conquest when a small group of Otomí Indians, presumably fleeing the Spanish invader, established a settlement near the mouth of the spring-fed canyon (later known as the Cañada) at the end of the long valley leading to Apaseo (Figure 4.1). The Otomí leader, Fernando de Tapia, soon allied himself with the Spaniards, but the pueblo emerged from obscurity only after the discovery of silver in Zacatecas in 1547. It lay on the main route to the mines at the border of unpacified Chichimec territory and became an important base for supplies and transportation.[1] In 1550, a Franciscan friar, Juan Sánchez de Alanis, joined Fernando de Tapia to move the pueblo a short distance to its present location and give it a proper Spanish form. It was laid out in the form of a chessboard on the southern bank of the stream that flowed from the Cañada. A canal conducted water from the stream to irrigate small gardens and fields that produced not only the indigenous staples of maize, chile, and beans, but a variety of European fruits and vegetables introduced by the Franciscans. Some wheat may also have been cultivated, but the evidence is conflicting.[2]

Swelled by immigration, the pueblo grew in the face of the general decline of the Indian population and required new allocations of water from the municipal canal. According to a seventeenth century tradition, Fernando de Tapia and his son and successor, Diego de Tapia, provided that the water

> divides at the dam, putting in the ditch the water gates mentioned in the sworn testimony...allocating to each barrio the quantity of water that each needs for its use and the irrigation of its gardens and cropland on the basis of the knowledge and experience of the said Don Fernando and Don Diego whom no one contradicted...[3]

The Indian population within the Cañada itself, which existed from the first days of settlement, could divert water directly to its fields from springs. Another Indian community was established in 1603 on the far side of the river from the town. With the authorization of the viceroy, the corregidor granted to the community an expanse of land that it irrigated from headgates located next to the town. A system for the allocation of water among the population seems to have been instituted with the original distribution of

---

[1]Jiménez Moreno, pp. 56, 74, 79, 84, 97; Gerhard (1972), p. 224; Powell, p. 25.

[2]Velázquez, I, pp. 15, 29-30, 39, 42-44; Ponce, I, p. 535.

[3]MNA.AM-Qro, roll 23, first doc., f. 277.

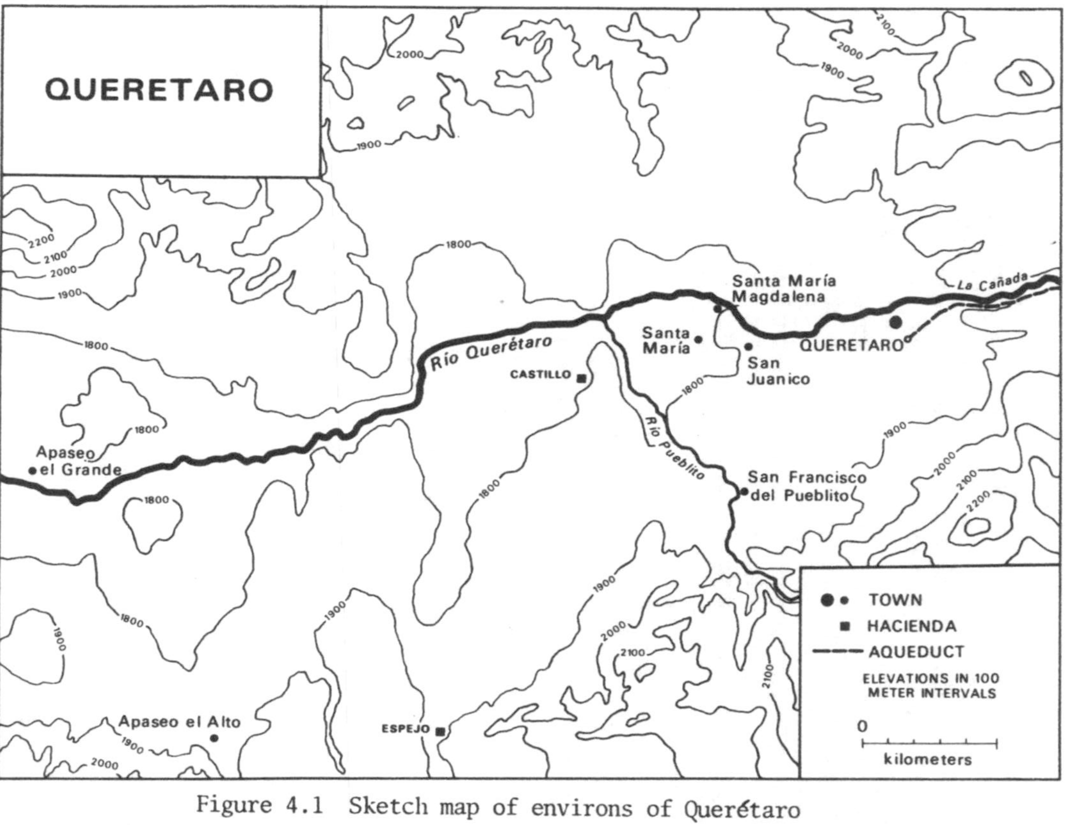

Figure 4.1 Sketch map of environs of Querétaro

land.[4] The area soon filled with houses, gardens, and milpas; it was known as the Congregación and later as the parish of San Sebastián.

The second cacique of Querétaro, Diego de Tapia, rose to extraordinary prominence for an Indian leader. It would be an understatement to say that Don Diego was well assimilated into Spanish society; he acquired wealth and social acceptance equaled by very few Spaniards. Through a remarkable display of business acumen, he acquired, mostly by purchases from Spaniards, a large personal estate with interests in mining, a house in Mexico City, cattle estancias, and an assortment of properties in and around Querétaro. His possessions were comparable to those of the richest Spaniards in the province. Don Diego frequently lent financial support to other members of his family, guaranteeing notes and assuming debts. With this assistance, three of his sisters, Doña María, Doña Beatriz, and Doña Magdalena also accumulated substantial properties.[5]

Toward the end of the sixteenth century, Don Diego began to put property to the west of the town--probably about ten caballerías--into wheat production. The part of this property later known as the Jacal Grande was irrigated from the municipal canal; other property north of the river, later called Santa María Primera, was irrigated from dams and canals that Don Diego himself built. The properties of other family members bordered both farms. The estates of Doña Magdalena and Doña Beatriz were later termed Santa María Segunda and Tercera, respectively, and that of Doña Maria, south of the river, was known as San Juanico.[6] All three were irrigated from the river. To serve these enterprises, Diego de Tapia secured a license in 1595 to operate a flour mill below an existing dam in the Cañada.[7] In 1599, he appears as the seller of 1,500 fanegas of wheat, the production of a medium-sized hacienda.[8]

A few Spaniards joined the Tapia family in establishing wheat farms in this period. In 1591, the encomendero, Pedro de Quesada, purchased from the Indian pueblo four caballerías with rights to water from the municipal canal;[9] the property later became a wheat farm known as the Hacienda Doña Melchora. The property of the Franciscan Capilla de San José near the end of the municipal canal was placed in wheat production at the beginning of the seventeenth century,[10] and two small Spanish wheat farms south of the town, Don Amaro and Callejas, may well also have had their origin at this time.[11]

In the early seventeenth century, Querétaro was thus bordered to the west by a fringe of wheat haciendas, while the land closer to town and in the Cañada was devoted to small gardens and milpas. The cultivation of the land was progressively intensified under the stimulus of population growth. The relación geográfica of 1582 reported that fifty

---

[4]MNA.AM-Qro, roll 23, first doc., f. 289; AGN-T, Vol. 417, Exp. 1, ff. 80v, 182; AGN-M, Vol. 24, f. 174.

[5]Super (1982), Chap. 10.

[6]Información de los Méritos, pp. 50, 61; AFC-CSC, leg. "títulos"; AGN-T, Vol. 417, Exp. 1, ff. 80v, 99; Vol. 2648, Exp. 2, ff. 33v-37v.

[7]AGN-M, Vol. 21, f. 73. See also Mercedes, Vol. 35, f. 33.

[8]Super (1982), Chap. 10.

[9]AFC-CSC, leg. "títulos."

[10]MNA.AM-Qro, roll 23, first doc., f. 310.

[11]AGN-T, Vol. 2648, Exp. 2, ff. 6-6v (Callejas and Don Amaro).

Spanish vecinos lived in the town.[12] After the cessation of Chichimec hostilities in 1590, there was a burst of construction of public and private buildings.[13] Felix de Espinosa asserted that the town had over 250 vecinos in 1604.[14] Vázquez de Espinosa, who visited Querétaro around 1620, reported that it then had 500 vecinos, a clearly exaggerated figure that nevertheless suggests further growth.[15] La Rea, a resident of Querétaro himself and probably well informed, states that it had almost 400 vecinos in 1639.[16] The earliest surviving report of the Indian population records 770 tributaries in 1643, for a total Indian population of perhaps 2,600.[17]

During these years, Querétaro emerged as the principal center of sheep raising in New Spain.[18] According to La Rea, the vecinos of Querétaro possessed over one million head of sheep in 1639.[19] These vast herds migrated seasonally in search of pasture along sheep walks extending west to the shores of Lake Chapala and north to the region of Monterrey.[20] Around 1660, about one-half of the wool produced in the province was sold to workshops in Querétaro itself mostly to be processed into rough woolen cloth.[21] In addition, the town's strategic location on the eastern border of the Bajío contributed to its rise as a provincial center of trade and a terminal for mule trains traveling to and from the capital. The province's exports to the capital were mostly derived from stock raising, but a substantial trade in wheat was also carried on.[22]

The Franciscan Convent of Santa Clara was the nexus of much of the social and economic life of the town in the seventeenth century. It was founded by Diego de Tapia when his only daughter desired to enter the religious life. Upon his death in 1614, Don Diego gave the convent most of his estate, including his wheat haciendas and flour mill.[23] His sisters were childless, and their property also passed to the convent, either by direct legacy or the execution of liens.[24] By mid-century, the convent had augmented these holdings by purchasing three small haciendas on the municipal canal, Doña

---

[12]Velázquez, I, p. 12.

[13]Super (1982), p. 12.

[14]Espinosa, p. 358.

[15]Vázquez de Espinosa, p. 135.

[16]La Rea, p. 283.

[17]Gerhard (1972), p. 225.

[18]Chevalier (1952), pp. 117-19, 238, 352.

[19]La Rea, p. 285. Sigüenza y Góngora gives the same figures in 1688. Sigüenza y Góngora, p. 2.

[20]Serrera Contreras, p. 293-94.

[21]Super (1982), Chap. 6; Super (1976), p. 212; Chevalier (1952), p. 136.

[22]Super (1982), Chap. 6; Florescano (1971), p. 90.

[23]AFC-CSC, leg. "títulos"; Super (1982), Chap. 10; La Rea, p. 256.

[24]AFC-CSC, leg. "títulos"; MNA.AM-Qro, roll 23, first doc., f. 293; Testamento de Doña Beatriz, pp. 475-89.

Melchora, Don Amaro, and Callejas,[25] and an impressive number of agricultural haciendas and stock raising estancias near Querétaro and in other parts of the Bajío. The revenues from these properties supported a large community within the town. In 1668, the Convent of Santa Clara sheltered almost 100 nuns and employed over 500 servants. The properties were later auctioned off in 1695 when the Franciscan administrators chose to rely on mortgage investments.[26] Between 1696 and 1720, the convent was a party to about eighty-five percent of the mortgages and capellanías recorded in Querétaro.[27]

Although surviving data are fragmentary, there can be little doubt that the convent expanded wheat production in its Querétaro haciendas in the forty years after the death of Diego de Tapia. The Hacienda San Juanico is reported to have harvested 1,540 fanegas in 1625 and 4,168 fanegas ten years later; in 1654, the convent was said to sow 680 fanegas in its Querétaro haciendas, suggesting a harvest in the order of 20,000 fanegas.[28] La Rea claimed that the wheat haciendas near Querétaro produced over 30,000 fanegas in 1639;[29] most were owned by the Convent. Its wheat production justified investment in an improved dam in 1634 to better serve the mill inherited from Tapia.[30]

It is still more difficult to document the intensified cultivation of Indian land, but, by all accounts, the Cañada was already densely settled by the mid-seventeenth century, and a village, San Pedro, emerged as a nucleus of settlement. The most important of the springs then irrigated the fields of twenty to twenty-five families. Although a little wheat was grown, the water was needed chiefly to support a growing population in a confined space. The Indians insisted that they could not support themselves without the use of the spring water.[31] At the western edge of the Congregación, an expanse of irrigated communal land was rented to Spaniards and became known as the Hacienda de la Comunidad. A cacique, Balthasar Martín, acquired a nearby plot of land, later known as the Hacienda San Juan and San Pablo, that he put into wheat production around 1630.[32]

The large majority of the Spanish landowners in the town possessed individual rights to water, and many cultivated gardens. In 1639, La Rea wrote,

[25]AFC-CSC, leg. "títulos." The property of Balthasar Martín was later known as the Hacienda San Juan and San Pablo and was held for a while by the Convent of Santa Clara. See Lavrin.

[26]Lavrin, passim.

[27]Super (1982), Appendix II.

[28]Lavrin, pp. 85, 97; MNA.AM-Qro, roll 23, first doc., f. 277v. The Haciendas San Juanico and Doña Melchora in 1668 had a yield of 25-30 to 1. MNA.AH, Fondo Franciscano, Vol. 92, ff. 93-4. For a general discussion of productivity, see Morin, pp. 238-42.

[29]La Rea, p. 284.

[30]MNA.AM-Qro, roll 23, first doc., ff.254v-63.

[31]Ibid, ff. 224-8, 278-8v.

[32]Ibid, ff. 239-44; AGN-T, Vol. 764, Exp. 3, Cuad. 1, f. 99; Vol. 868, Exp. 6.

> Surrounding the city there is not a foot of land that is not cultivated with seeds of all kinds, beautiful gardens, large vineyards that yield a good grape harvest, together with all sorts of Castillian fruit trees.[33]

The record of land compositions in 1643 gives solid corroboration to this panegyric: no fewer than forty-nine gardens are mentioned in an assessment list of urban properties.[34]

The complex distribution of water needed to serve this patchwork of wheat haciendas, Indian communities, house sites, and small horticultural plots evolved in an incremental fashion in the first half of the seventeenth century. Individual grants of water rights were rare,[35] but there was a progressive parcelization of irrigated land with appurtenant water rights. Diego de Tapia, as cacique, sometimes sanctioned the conveyance of irrigated communal land.[36] Water disputes were resolved in a practical manner by orders that defined the size of water intakes and the times of irrigation turns. For example, when a Spaniard sought to secure an irrigation turn on the canal of the Congregación, the alcalde mayor awarded him a turn of one-half an hour at unstated intervals; on appeal he obtained a turn of one hour every third day.[37] Some time before 1650, the alcalde mayor made a general assignment of irrigation turns for the wheat haciendas west of town;[38] other similar allocations may have been elsewhere. The enforcement of such orders seems to have rested on the normal municipal officials; one finds no mention of officials specially charged with responsibility for water.

There were important improvements in the water supply system in the first half of the seventeenth century. Around 1640, the Convent of Santa Clara built a dam deep within the Cañada to store water during the summer that could be released when needed in the dry season. The spring water was thus supplemented by surface waters from the relatively small drainage basin above the Cañada. The convent's license to construct the dam provided that the increased supply of water was to be allocated exclusively to its wheat haciendas. The cost of the dam, paid by the convent, was about 7,000 pesos (Figure 4.2).[39]

The municipal government also directed some investments in the water supply. Witnesses at mid-century allude to the recent construction of a masonry dam that replaced the former wooden dam used in the distribution of water. The project cost 3,000 pesos, which was assessed among the residents of the town on a pro rata basis. The gates of this municipal dam released sixty percent of the water to the municipal canal and forty percent to the Congregación and the wheat farms on the river.[40]

---

[33]La Rea, pp. 283-84.

[34]MNA.AM-Qro, roll 23, first doc., f. 92 et seq.

[35]I have found only two grants of water rights in the seventeenth century. AGN-M, Vol. 28, f. 14v; Vol. 35, ff. 143v, 186.

[36]E.g., AFC-CSC, leg. "títulos"; MNA.AM-Qro, roll 5 (Nicolás de Robles, December 9, 1605); AGN-M, Vol. 28, f. 14v.

[37]AGN-T, Vol. 2785, Exp. 19; also AGN-M, Vol. 50, f. 3v.

[38]MNA.AM-Qro, roll 23, first doc., ff. 227v, 278.

[39]AFC-CSC, leg. "títulos"; AGN-T, Vol. 2785, Exp. 17; AGN-M, Vol. 41, f. 46v.

[40]MNA.AM-Qro, roll 23, first doc., ff. 278v, 318; AGN-M, Vol. 50, f. 3v.

When the Convent of Santa Clara attempted to use its newly constructed dam to expand irrigation of its haciendas, it aroused the opposition of Indians who had expected to benefit from the project. The Indians' petition to the audiencia evidently converged with those of several other parties who had grievances of their own. In 1654, the audiencia dispatched oidor Gaspar Fernández de Castro to make a general adjudication of water rights in the town.[41]

The entire text of Fernández de Castro's ruling survives;[42] it includes seventy-six pages of detailed regulation of water rights quite free of redundant legal verbiage. One is immediately struck by its complexity. The oidor regulated the distribution of water from sixteen principal water gates, established 107 turns of water usage (most for very short periods of time) and defined the rights of over 200 Spaniards as well as those of the wheat haciendas and the many groups of Indians within and around the town. He did not, however, devise this extraordinarily elaborate system; he attempted merely to restate and precisely define the rights that had evolved through innumerable practical adjustments. In only two cases did he explicitly increase or reduce existing water rights.[43]

The sizing of irrigation gates and other water intakes was the principal means for the allocation of water. At the municipal dam, there was a three-part division: two surcos were channeled to the Congregación; six surcos to the river for the benefit of downstream wheat haciendas; and twelve surcos to the principal canal of the town. Three headgates on the canal tapped a full surco: one was directed principally to the two wheat farms south of town, Don Amargo and Callejas, and the other two led with a few diversions to the haciendas of Jacal Grande and the Capilla de San José. All other water intakes were of remarkably small size; twenty-one were of one-half a naranja; thirteen were measured in terms of reales. We will discuss the significance of these measures in a later chapter; for the moment, they may be accepted at face value--a naranja was a flow comparable to a coin. The conduction of such small quantities could only be accomplished with the aid of masonry or clay conduits whose existence is confirmed by the many references to alcantarillas, cañerías, and canales de piedra.

Although some parties secured a continuous supply of water, most users had a right only to a periodic turn. Some turns were for periods as limited as one hour every other week. A quotation of the allocation for a particular water intake may help to convey something of the texture of the system:

> And on the corner where Antonio de la Parra has customarily placed a ditch dam to irrigate his garden the ditch shall be raised and narrowed so that by this means he may irrigate through an intake of one naranja Monday and Thursday of each week from six in the morning to six in the evening so that the water goes to the canal and to the fields that irrigate with the excess water under penalty of twenty pesos for each violation and through the said intake Tuesday of each week Juan Martín may irrigate from six to eight in the morning; Francisco Martín de Lorca from eight to twelve; and Melchor de los Reyes from twelve to four in the afternoon.[44]

---

[41]AGN-T, Vol. 868, Exp. 6; MNA.AM-Qro, roll 23, first doc., ff. 224-323.

[42]AGN-T, Vol. 2648, Exp. 2.

[43]Ibid, ff. 22v, 23v, 49v-60v.

[44]Ibid, f. 33.

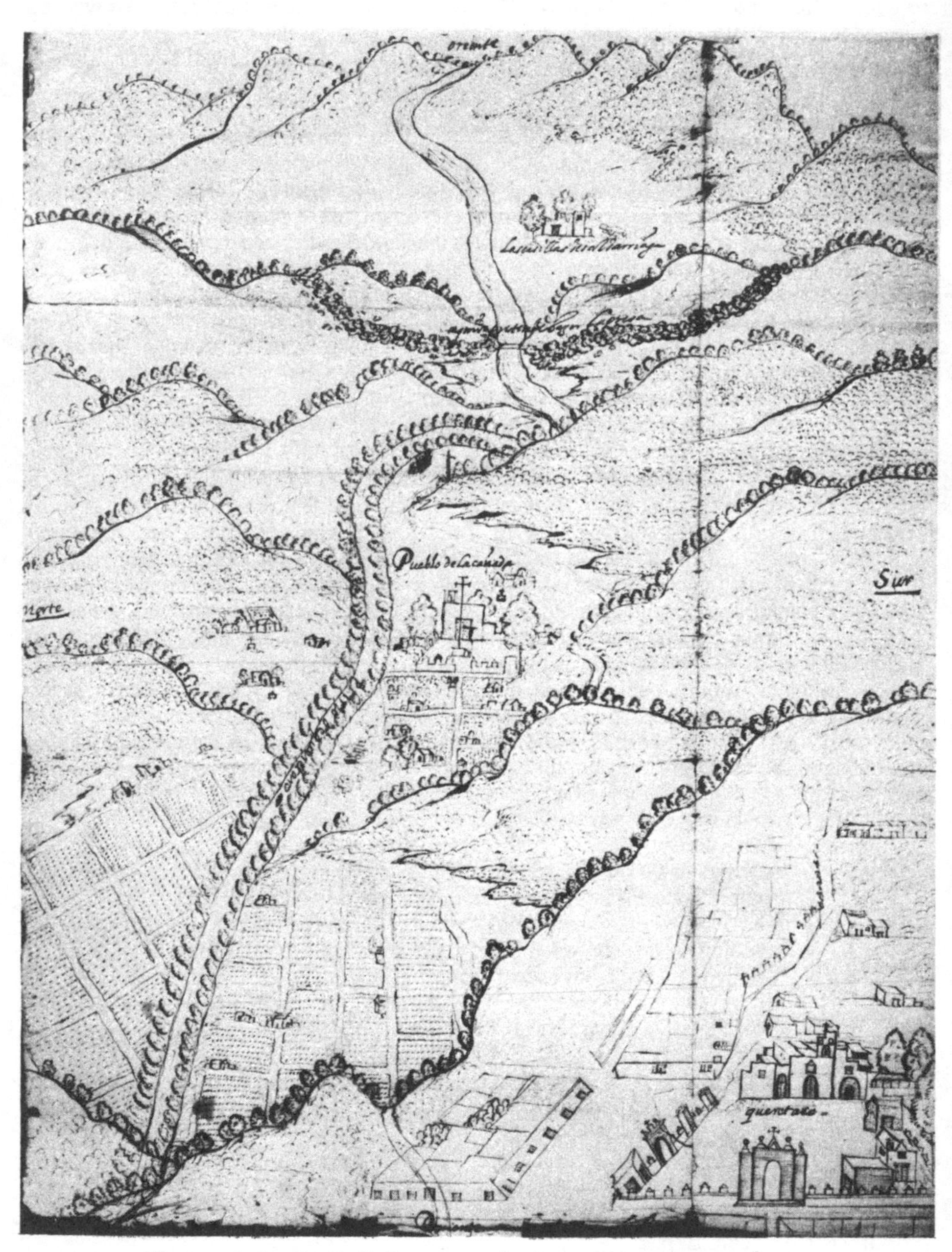

Figure 4.2 Dam of Convento de Santa Clara, Querétaro (Source: AGN-T, Vol. 2785, exp.13)

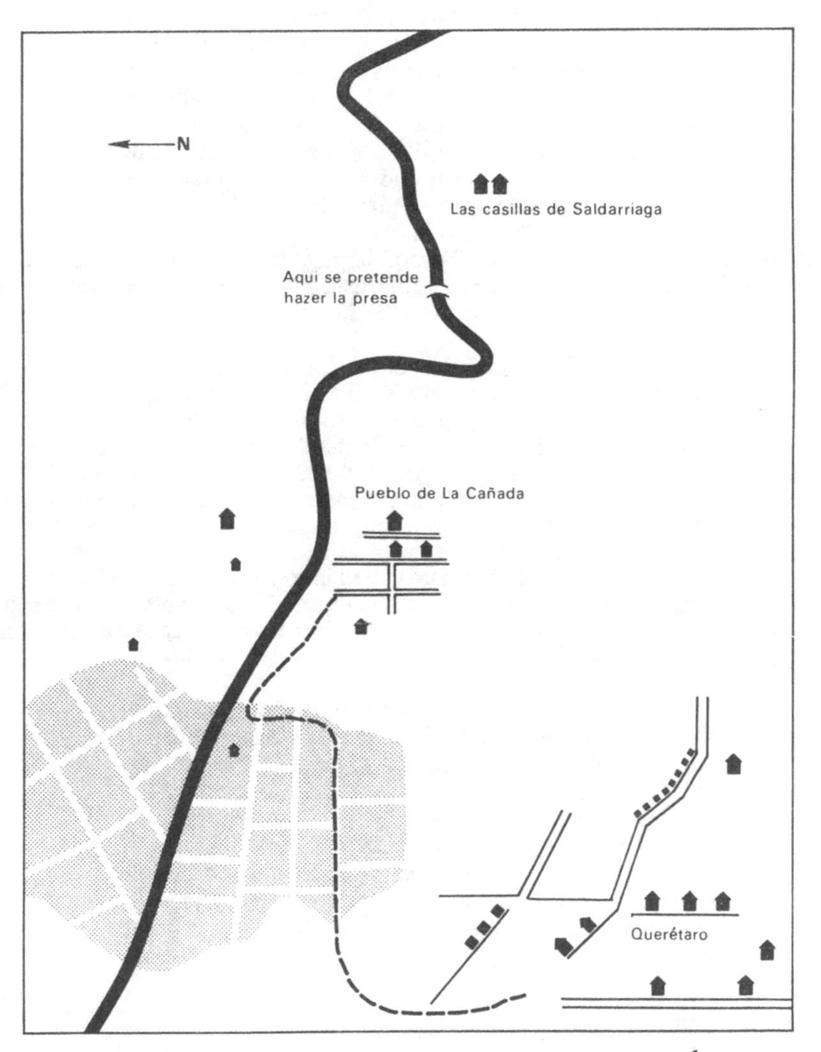

Figure 4.2 Dam of Convento de Santa Clara, Querétaro

Many intakes were fitted with water gates and keys, and the order frequently prescribed that the gates by locked after the turns.

The oidor's order provides further evidence of the abundance of gardens. About two-thirds of the Spanish residents mentioned in the order were given the right to water gardens, ranging from small jardines within the domestic enclosure to numerous huertas grandes serving commercial purposes. We can believe Sigüenza y Góngora when, like La Rea earlier in the century, he described Querétaro after a visit in 1689 as follows:

> The effect of the dry climate is mitigated by the ample quantity of water that irrigates and refreshes the location and surroundings of the city...There is no house, however small it may be that does not have access to water that comes from wells or is conducted by masonry troughs that pass through the streets of the city...As a consequence of these facilities and the fertility of the soil, one finds everywhere delicious gardens and pleasing and productive horticultural plots that have no equal in Mexico.[45]

Querétaro was indeed a city of gardens! Several large gardens lay at the eastern edge of the town near the Cañada, but the order reveals that, like the early medieval towns of Europe, the patchwork of gardens was interwoven into the fabric of every section of the town.[46]

The order gave the Indian population in the Cañada exclusive right to specified springs that were to be enclosed with masonry walls to form reservoirs, and it prescribed the posting of watchmen to ward off invasions of livestock. The two surcos assigned to the Congregación could be shared by only two Spaniards who had rights to limited and periodic supplies. Within the Congregación itself, the allocation of water was evidently governed by custom since the order is silent on the matter. The allocation of water rights to Indians in the town of Querétaro reveals a very dispersed distribution of the Indian population, as a consequence of the town's origin as an Indian pueblo. Areas described as barrios were small and ill-defined; Spaniards lived in so-called barrios, while groups of Indians lived in areas of predominantly Spanish residence. Water rights were often given collectively to groups of Indians but sometimes assigned to individually named Indians. For example, the order provided that

> In the house and garden of Joseph de Breña an intake of half naranja be placed with which he may irrigate Friday of each week for two hours from six to eight in the morning, and Sebastian de Santiago, Nicolás Gonzales, Luiza Juliana and María de la Cruz, Indians, may successively irrigate on that day for one hour their gardens that are behind that of said Joseph de Breña, and Cathalina de Bargas and Bartola de Vargas, Venito García and Juan Ximenes, Spaniards, may irrigate on that day another four hours in the order mentioned.[47]

---

[45]Sigüenza y Góngora, pp. 3-4.

[46]Compare Mumford, p. 288.

[47]AGN-T, Vol. 2648, f. 16v.

A large segment of water users were given only rights to remanientes. These residual rights were assigned not only to innumerable small plots within the town, but to four wheat farms at the outskirts. The cacique, Balthasar Martín, possessed a right to the remaniente of the two surcos assigned to the Congregación, and three haciendas--Jacal Grande, Doña Melchora, and Capilla de San José--had a right to irrigate from excess waters from the municipal canal. Each was allocated three separate turns of twenty days of water from the canal during the winter wheat season. The right to remanientes was largely a right to nocturnal irrigation. The water turns assigned to the town were without exception for daytime hours, and during the dry season they must have consumed nearly all the water from most branch canals.

The flow of the river was distributed among wheat haciendas by irrigation turns: the three haciendas named Santa María each received twelve days of water; the Hacienda San Juanico, sixteen and one-half days; and the farm of Balthasar Martín and another small property, a total of three and one-half days. These sixty-eight days of irrigation were rotated among the haciendas three times during each winter. Beginning on May 12 of each year, there began a completely new regime, known as the media tanda. The maize crop, vital to the Indian population, had then been sown. To assure successful germination of the crop before the intense summer rains, the Indians were given an increased share of the water: the river was divided into equal shares between the Indian community on the north bank and the wheat haciendas. The latter, in turn, allocated the water for maize cultivation by reducing the winter irrigation turns by one-half; the Hacienda Santa María Primera, for example, received six days of water on three occasions during the summer. A similar arrangement was worked out between the haciendas irrigating from the municipal canal and neighboring groups of Indians.[48]

Responsibility for enforcement of the order lay with the normal municipal authorities, but the alcalde mayor was specifically given custody of the keys to the many water intakes with locked gates. Some designated agent of the alcalde must have intervened in the daily operation of the system by opening and closing the water intakes.[49] Outside the town, the system depended on self-enforcement, but the order prohibited water guards from carrying guns; only swords and daggers were allowed. The responsibility for cleaning the major common portion of the canal was conveniently placed on the three haciendas receiving the remanientes. Other private parties were expected to maintain the conduits directly serving them. The haciendas and the Indian communities were required annually to pay small fees of 179 pesos each for the maintenance of the system.[50]

The oidor's order was aimed not only at defining water rights, but at putting the infrastructure in order. An architect, Francisco de Echavida, was directed to construct water intakes according to a uniform standard, raise the municipal dam by one-third, replace irrigation gates, build nine main conduits, and carry out a variety of other specified improvements. The order contains elaborate provisions for assessing these costs and the fees of Fernández de Castro and Echavida among all the benefited parties. The Convent of Santa Clara was assessed one-third of the cost of repairing the dam, forty percent of Echavida's salary, and almost one-half of the oidor's salary. Fernández de Castro remained in Querétaro 124 days and charged twenty-nine pesos and five tomines for each day--a handsome sum indeed![51]

---

[48]Ibid, ff. 38v, 62-63v.

[49]Ibid, f. 42v.

[50]Ibid, ff. 38-42v.

[51]Ibid, ff. 65-75.

The city of Querétaro grew in the last 150 years of the colonial period to be one of the largest cities in the Western Hemisphere. In 1742, Villaseñor y Sánchez reported a population of 2,805 Indian families and 3,004 families of Spaniards, mulatos, and mestizos; the total population thus may have been about 24,000.[52] Toward the end of the colonial period, one finds several population estimates, each presenting peculiar problems of interpretation. A census of the archbishopric of Mexico in 1779 has the virtue that it clearly indicated the areas included in the count; it reported a population of 23,848 persons for the parish of Santiago, the main part of the city south of the river, and a population of 13,136 for the parish of San Sebastian, the predominantly Indian congregación north of the river.[53] There is evidence that the population continued to grow in the last decades of the colonial period. Super concludes that "it is safe to say that the city at its largest before 1810 had close to 50,000 inhabitants."[54]

At the end of the colonial period, Querétaro was a remarkably industrialized city with an economy based largely on textiles and tobacco processing. Around 1800, its brilliant corregidor, Miguel Domínguez, stated that 9-10,000 persons worked in the textile industry and 3,000 persons in the Royal Tobacco Factory.[55] Together the two industries accounted for a high percentage of the economically active population, but the city retained its importance in regional trade, and, with its many amenities, became a social refuge for the upper classes. Viceroy Branciforte described it in 1795 as "a city of rich merchants and wealthy hacendados."[56] It evolved an increasingly complex and impressive religious establishment; in addition to the Convent of Santa Clara, the Franciscans founded the College of the Holy Cross that was of great importance in directing missionary activity in northern Mexico (Figure 4.3).

In the absence of tithe data, one cannot know precisely the importance of irrigated wheat production in the late colonial period, although the construction of new flour mills suggests a substantial increase.[57] But nineteenth century data give us some insight. Querétaro continued to export wheat to the capital, but in the economy of the province as a whole, wheat production had a distinctly secondary importance. In 1840, the state of Querétaro produced 624,000 fanegas of maize and 43,720 fanegas of wheat.[58] Much of the wheat production was centered in the valley of Querétaro, but some was grown in San Juan del Río, the valley Tequisquiapan, and the municipality of Santa Rosa.[59]

The pattern of irrigation around Querétaro established by Fernández de Castro's ruling remained recognizable throughout the colonial period despite some consolidation of properties. Six haciendas irrigated from the river west of Querétaro in 1794--the same number as in 1654--and only one new name appears on the list. The haciendas

---

[52]Villaseñor y Sánchez, I, Book 1, p. 93.

[53]Serrera Contreras (1973), p. 541.

[54]Super (1982), Chap. 1.

[55]Serrera Contreras (1973), pp. 496, 502.

[56]Ibid, p. 503. See generally, Super (1982).

[57]There were two flour mills in 1767 and four in 1794. AGN-M, Vol. 81, f. 235v; AGN, Industria y Comercia, Vol. 32, Exp. 3, f. 55v; AGN-H, Vol. 578B, leg. 1, f. 139v.

[58]Raso, p. 38.

[59]Septien y Villaseñor, pp. 202-4. The Hacienda Carrillo was one of the Haciendas Santa Maria. AGN-T, Vol. 2738, Exp. 38, f. 18.

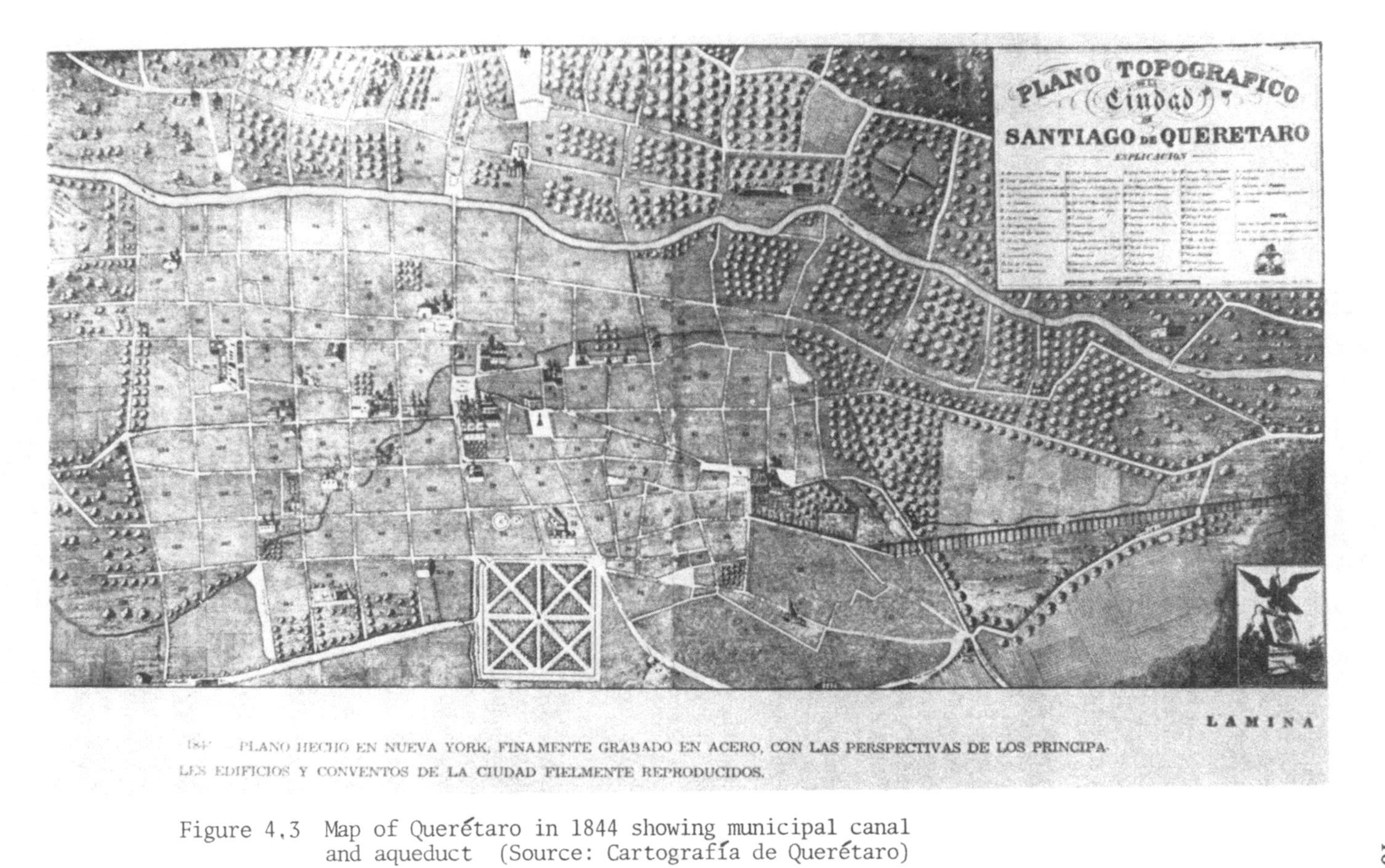

Figure 4.3 Map of Querétaro in 1844 showing municipal canal and aqueduct (Source: Cartografía de Querétaro)

collectively seem to have received the same proportion of river water as before.[60] The excess water from the municipal canal retained its value despite the increased population of the town; in 1713, the twenty days of water of the Hacienda Doña Melchora was appraised to be worth more than the value of the land.[61] This hacienda was evidently absorbed into one of its neighbors; in 1794, only the Haciendas Jacal Grande and Capilla de San José received remanientes from the canal.[62]

The channeling of water to storage basins, however, acquired an importance that was not foreseen by the oidor's ruling. Despite its extremely detailed description of rights, the order did not assign irrigation turns to the wheat haciendas for brief periods between the wheat and maize seasons. From at least as early as 1660, the flow of the river was in fact allocated equally among haciendas to fill ponds between May 2 to 12.[63] The flow during the period October 3 to 10 was eventually preempted by a single landowner, presumably also to fill storage basins. In the only modification of Fernández de Castro's ruling that I have found, the owner of the Hacienda de la Comunidad obtained from the audiencia in 1746 a grant of seven days of water during this period, and despite opposition, actually enjoyed use of the right conferred by the grant.[64] The haciendas on the municipal canal also stored water when it was not used for irrigation; the inventory of the Hacienda Doña Melchora in 1713 describes "a large pond to store water."[65]

The many water disputes in the later colonial period always referred to the oidor's order as the guiding authority.[66] It remained the only comprehensive plan for allocation of water despite the growth and transformation of the city. Newly established obrajes encountered vigorous opposition when they attempted to appropriate water rights originally allocated to others.[67] Indian barrios, always under pressure, invoked the order to defend their interests. In various lawsuits, they sought to compel other users to repair or install proper water intakes, disputed the rights of others to receive water assigned to the barrios, and litigated the appropriate irrigation turn for water intakes.[68] The holders of rights to remanientes struggled to maintain their share in the face of increased water consumption. The owner of the Hacienda San Juan and San Pablo held the right to the remaniente of the two surcos assigned to the Congregación, but in 1758, there was little water left for the hacienda; 740 families then lived in the area of the Congregación. The

---

[60]AGN-H, Vol. 578B, leg. 1, f. 139v (refers to five haciendas, apparently lumping together the two Santa Marías); AGN-A, Vol. 138, Cuad. 8, f. 78v (mentions a seventh hacienda, San Pablo, receiving a very small share of water). With regard to the share of water, note reference to division of twenty surcos in AGN-H, Vol. 578V, leg. 1; and see AGN-T, Vol. 2738, Exp. 38, f. 6 (1746); Vol. 2967, Exp. 112, f. 298 (1762).

[61]MNA.AM-Qro, roll 51 (Diego Antonio de la Parra, June 26, 1713).

[62]AGN-H, Vol. 578B, leg. 1, f. 139v.

[63]AGN-T, Vol. 417, Exp. 1, ff. 198v; Vol. 2738, Exp. 38, f. 21.

[64]AGN-T, Vol. 2738, Exp. 38, ff. 8v-34v; AGN-M, Vol. 75, f. 102.

[65]MNA.AM-Qro, roll 51 (Diego Antonio de la Parra, June 26, 1713).

[66]E.g., AGN-T, Vol. 417, Exp. 1, ff. 204-8; Vol. 1970, Exp. 107; Vol. 2765, Exp. 16.

[67]AGN-T, Vol. 455, Exp. 4; Vol. 764, Exp. 3, Cuad. 3, f. 4; Vol. 2765, Exp. 11.

[68]MNA.AM-Qro, roll 34, Arch. Not. 4 Qro, leg. civ 1705-1709 and leg. civ 1688-1694 (barrios San Sebastián, San Roque, and San Francisco).

owner attempted to negotiate a new arrangement giving him a right to a nocturnal water supply, but in the ensuing litigation, he was relegated to his right to a then nonexistent remaniente.[69]

The water from the Cañada continued to irrigate a profusion of gardens. A description of the city in 1794 mentions a "league of gardens" along the river at the mouth of the Cañada.[70] The only surviving tithe record, that of the *diezmo chico* of 1732, contains a list of 372 Indian gardens whose owners had paid the tithe.[71] Eleven barrios were included in the assessment; the largest concentrations of gardens were in the Cañada, the barrio San Isidro near the entrance to the Cañada, the barrios San Sebastián and San Roque, north of the river, and the barrio Espíritu Santo at the southwestern edge of the city.[72] A few are clearly depicted in maps of the city dating from the late eighteenth century. But we may reasonably doubt whether much of the patchwork of urban gardens within the city survived. At the close of the colonial period, the city had only 155 blocks, and the residential density was very high, perhaps in the order of 250-300 persons per block. There could not have been much room for gardens.[73]

As early as the seventeenth century, the water of the municipal canal suffered contamination. Fernández de Castro forbade a slaughterhouse from draining wastes into the canal.[74] But the most serious pollutors were the numerous tanneries and obrajes. In 1721, the corregidor of Querétaro complained of the "bad water" that caused "grave illnesses" among the population and laid the blame on the obrajes for "spoiling (*viciando*) the water that passed through them."[75] The obrajes deposited into the canal, among other noxious substances, *alcaparrosa*, a ferrous sulfate used as a dye.[76] The owners of the wheat haciendas receiving remanientes complained on occasion that the water had become unfit for irrigation.[77] A Jesuit priest objected that the clothes of sick persons were being washed in the canal, a complaint that reflected a justified concern for the healthfulness of the water.[78]

It is not surprising that much of the population in the early eighteenth century relied on water carriers for potable water. A spring at the mouth of the Cañada remained uncontaminated and was close enough that water carriers could make several trips a day

---

[69]AGN-T, Vol. 764, Exp. 3, Cuad. 1, f. 4v, 71-126, Cuad. 3, f. 70, 80, Cuad. 5, f. 10; Vol. 2967, Exp. 112, f. 298.

[70]AGN-H, Vol. 578B, leg. 1, f. 139v.

[71]CAAM, Book 374.

[72]Super (1982), Chap. 2. See AGN-T, Vol. 455, Exp. 4; Vol. 764, Exp. 3, Cuad. 3, f. 10; MNA.AM-Qro, roll 34, Arch Not 4, Qro, leg. civ 1688-1694 (barrio Santa Ana).

[73]*Cartografía de Querétaro*, laminas I-V, XXV.

[74]AGN-T, Vol. 2648, f. 28v.

[75]AGN-M, Vol. 70, f. 105v et seq. Also AGN-M, Vol. 71, f. 310-17; AGN-T, Vol. 2765, Exp. 11, f. 1v.

[76]Super, Chap. 2.

[77]AGN-T, Vol. 455, Exp. 4, f. 1v; Vol. 764, Exp. 3, f. 25v.

[78]AGN-T, Vol. 2765, Exp. 11, f. 1.

to the city.[79] But this laborious method of distribution did not satisfy the needs of a growing population.

On November 5, 1721, the corregidor of Querétaro commissioned three experts, including an architect, Diego de Andizabal y Sarate, to study the feasibility of bringing water to the city from a spring located about seven kilometers away in the Cañada. The experts proposed the construction of a conduit along the side of the Cañada that would lead to an aqueduct spanning the low-lying land to the east of the city, and would terminate at the plaza of San Francisco, the highest point in the city (Figure 4.3). The cabildo promptly secured the permission of the viceroy to pursue the project. The estimated cost was first put at 20,000 pesos and then raised to 25,000 pesos. The viceroy directed the city to raise a major portion of this amount from assessments on the offending obrajes "which ought to make the largest contribution"; the balance was to be collected primarily from a general assessment imposed on all residents of the city.[80]

The project may have been instigated by a wealthy resident of Mexico City, Juan Antonio de Urrutia y Arana, Marqués de la Villa de Villar de Aguila. The Marqués was a patron of the Convent of Capuchin nuns, which established a house in Querétaro on July 21, 1721. He accompanied the nuns to Querétaro and soon bought a large home to permit annual visits.[81] We are told by Navarrete, writing in 1739, that, when the abbess impressed upon the Marqués the need for a supply of pure water, he offered to build the system at his own expense on condition that the city contribute 25,000 pesos. The offer, according to Navarrete, was sealed by a written contract with the cabildo.[82] The Marqués was indeed commissioned by the viceroy in 1723 to collect the 25,000 pesos assessed on the obrajes and residents of the city,[83] and he devoted himself with singular zeal to the project during the next fifteen years of his life.

The Marqués carried out a census of the population and prepared a detailed assessment of residents that filled 200 folios; 3,000 pesos were assessed against religious bodies and 22,000 pesos against individual residents and obrajes. Another 1,619 pesos was levied to cover administrative costs. Nearly all of this sum had been collected by 1728. The cabildo was unwilling, at least initially, to help finance the project, and municipal revenues were in any event only about 5,000 pesos a year. The Marqués tried unsuccessfully to secure the 179 pesos levied annually for the maintenance of the municipal canal.[84] The construction could proceed only because the Marqués generously contributed his personal funds.

Construction of a masonry reservoir at the location of the spring began in 1726 and was completed two years later.[85] The conduit itself took seven years to build. Throughout this time the Marqués was in close contact with the viceroy, reporting

---

[79]Ibid, ff. 8, 12.

[80]AGN-M, Vol. 70, f. 105v; Vol. 71, f. 310.

[81]Ramírez, pp. 76-78.

[82]Francisco Navarrete, pp. 63-67.

[83]Samaniego, p. 13; AGN-M, Vol. 71, ff. 317-8.

[84]AGN-M, Vol. 72, ff. 75, 79v; Samaniego, p. 14.

[85]Gacetas de México, I, p. 75; Francisco Navarrete, pp. 68, 73.

progress and securing necessary authorizations.[86] Navarrete has given us a colorful description of the work.[87] A masonry conduit, built in the Cañada, traversed tortuous terrain and extended underground for about one-half its length:

> At times along the steep slopes, then over the deep chasms;
> at times appearing above arches to traverse large barrancas;
> then disappearing entirely from view.

The most heroic stage of the project, however, consisted of constructing the aqueduct leading from the side of the Cañada to the city itself. Massive foundations were built to assure the stability of the aqueduct.

> It was necessary to escavate [sic] to such a depth to secure a solid and firm foundation that the hills and quarries long groaned to find themselves depleted and dispossessed of millions of fanegas of lime that was deposited forever in these deep crypts. Five varas on the side, twenty varas square and fourteen deep, the foundations were so massive that one could erect on them arches of surprising grandeur and height.

The scaffolding to build the aqueduct was itself an impressive construction.

> To build the masonry arches, corresponding to the height and majesty of the pilars [sic], it was necessary to transport entire forests in the form of planks, timbers and beams to the valley in order to build the scaffolding required for such a heavy and elevated structure; considering also the myriad of winches to raise the materials, pulleys, ropes, knots, bindings, leather thongs, hods, boxes and other equipment, it is easier to omit their description than to confound the memory with the prodigious number.

It was one of the greatest aqueducts in Spanish America when completed in 1735 (Figure 4.4). The wall of the reservoir was two and one-half meters high and 167 meters in circumference. The aqueduct proper was 1,280 meters long with seventy-four arches resting on deep foundations. At its highest point, the aqueduct was 28.4 meters above the ground. This imposing structure supported a relatively small conduit, about one-third of a meter wide and one-half a meter high. It was said to provide five surcos of water, which was one-fourth the flow of the river and less than one-half that of the municipal canal.[88] The aqueduct, however, was not intended to cover the needs of irrigation but to provide a domestic supply.

The final cost of the work exceeded the original estimate by a factor of about five and was financed principally from the personal fortune of the Marqués. The figures of Navarrete are corroborated by two other contemporary sources: the residents of the city

---

[86]AGN-M, Vol. 72, ff. 75, 79v; MNA.AM-Qro, roll 27, Mercedes de Agua; Samaniego, p. 14.

[87]Francisco Navarrete, p. 70.

[88]For physical description of the aqueduct, see Septien y Villaseñor, pp. 176-77; Romero de Terreros, pp. 61-62.

Figure 4.4 The Querétaro aqueduct in 1981

contributed 24,504 pesos; an anonymous donor gave 3,000 pesos; a condonación yielded 2,300 pesos; municipal revenues and the sale of water rights produced 12,000 pesos; and the Marqués gave the extraordinary sum of 82,986 pesos. The total costs, according to Navarrete, was 124,791 pesos.[89]

The plans called for the construction of a reservoir near the Plaza San Francisco from which the water would be conducted by underground pipes or conduits to public fountains, convents, and homes.[90] This phase of the project took another three years to complete and seems to have been financed by the sale of water rights. The viceroy authorized the Marqués to grant fully valid water rights, without need of further confirmation by the audiencia, through a Commission of Pure Water (Comisión de Aguas Limpias) over which he presided.[91] On October 17, 1738, water at last flowed to ten public fountains, all the convents of the city, the government buildings (casas reales), the jail, and numerous private homes.[92]

When Navarrete wrote his eulogy of the water system in 1739, the Commission of Pure Water had issued sixty grants of water rights to private homes and was in the process of issuing more. We have a record of two early grants made in 1737; both conferred the right to draw remanientes from the fountain in the Plaza Mayor. The practice of granting rights to the remanientes of public fountains continued to be common. A person willing to extend a water line and build a new fountain could expect to secure a grant of rights to the remanientes as compensation for his service. Thus, a priest, who built an especially lengthy extension of a line, was given a right to nine pajas from the trough of the newly constructed fountain. In this way, private investment was enlisted for expansion of the water service, but the needs of the public were given a certain priority.[93] The new fountains served more than utilitarian purposes. Some were of high artistic value, and they attracted social gatherings of townspeople. A rather puritanical citizen pleaded for a new extension of a water line on the grounds that the existing fountain encouraged contacts between young men and women.

Throughout the remainder of the colonial period, the water was distributed in smaller and smaller flows to an increasing population. In the decade of 1750, the characteristic grants was two pajas and sold for 150 pesos; at the end of the colonial period, it was only one paja but sold for 250 pesos. Loss of water in conduction had increased despite periodic repairs of the aqueduct and urban conduits. In 1800, it was reported that "there had been a notable shortage in almost all public and private fountains of the city."[94] The city then possessed twenty-two public fountains, but many more were

---

[89]Francisco Navarrete, p. 77. The Gacetas de México in 1738 gives the same figures for the Marqués' contribution and the same total. Gacetas de México, III, p. 162. Villaseñor y Sánchez gives the total cost as 114,000 pesos. Villaseñor y Sánchez, p. 90.

[90]Gacetas de México, I, p. 75.

[91]MNA.AM-Qro, roll 27, Mercedes de Aguas.

[92]Francisco Navarrete, p. 76.

[93]This paragraph and much of the material in the next two paragraphs is drawn from MNA.AM-Qro, roll 27, Mercedes de Agua. For a wealth of information on the fountains, see Septien y Septien, passim.

[94]AGN-H, Vol. 439, Exp. 19, f. 9.

needed.[95] It was said that "one does not hear of anything else among the poor except the complaint of water shortage."[96] But the scarcity of supply did not deter the cabildo, in need of funds, from issuing more water rights at higher prices.

A water right permitted the holder to secure a continuous flow through a connection installed at his own expense from a fountain, water main, or another house. Rights could be transferred only with the permission of the municipality. The water commonly flowed through underground clay pipes, but metal connections were also used.[97] Ordinarily, water passed through two or three houses before being discharged into the municipal canal. The second and third houses possessed rights to "remanientes de segunda (or tercera) concesion" that conferred the privilege of using water draining from the previous house. The house last in line was invariably required to pay the cost of conducting excess water to the municipal canal. The water from the canal itself maintained its importance in irrigating gardens and peripheral farms.[98]

The city retained a Commission of Pure Water and a water judge during the remainder of the colonial period. The judge of whom we have record held other municipal offices, such as regidor or alguacil mayor, and received only a small stipend (sixty-three pesos in 1783)[99] for his duties in this office. The principal responsibility of the water judge seems to have been to review requests for grants of water from the municipal canal.[100] The city also employed one or two fontaneros,[101] who were charged with

> taking care that the public does not lack sufficient water for its daily use...[and] that the pipes and conduits do not silt up or cause contamination of the water as it passes to its assigned outlet.

One finds little information regarding the administration of water within Indian communities, but a few brief passages reveal informal and variable arrangements without clearly specified turns or rights. For example, the Indians in the barrio San Sebastián were said to use the water "whenever they need it without having fixed days or assigned hours."[102] The barrio, however, had elected water administrators (repartidores electos) in 1798.[103]

---

[95]Zeláa e Hidalgo (1803), p. 9. The corregidor had estimated in 1790 that the city needed from thirty-three to thirty-eight fountains. See Super, Chap. 2.

[96]AGN-H, Vol. 439, Exp. 19, f. 1.

[97]Bribiesca Castrejón, passim.

[98]AGN-H, Vol. 439, Exp. 19, f. 1.

[99]AGN-A, Vol. 138, Cuad. 8, e.g., f. 80.

[100]MNA.AM, Colección Ant. T-2, 1 (Actas de Cabildo, 1811); MNA.AM-Qro, roll 27, Mercedes de Aguas; AGN-A, Vol. 147, Cuad. 1, ff. 4-5.

[101]AGN-A, Vol. 225, Exp. 2, f. 2.

[102]AGN-T, Vol. 764, Exp. 3, Cuad. 1, f. 33, Cuad. 3, f. 10; MNA.AM-Qro, roll 34, Arch Not 4 leg. civ. 1688-1694.

[103]AGN-T, Vol. 1296, Exp. 3, f. 3.

Every year the municipal government cleaned the aqueduct and reservoirs and made certain repairs and improvements to the system, but the meagre municipal revenues (only 7,255 pesos in 1780) did not permit adequate maintenance.[104] In the sample of records I have studied, the annual repairs usually amounted to less than 100 pesos.[105] In 1800, a report recommended repairs to the aqueduct estimated to cost 12,000 pesos, and two years later an expenditure of 3,252 pesos was actually authorized.[106]

The municipal ordinances contained only two provisions relating to water, both concerned the maintenance of the municipal canal.[107] The public was enjoined not to throw garbage or horse manure into the canals or to make unauthorized diversions of water. In addition, the users were required to clean and repair the ditches that directly served them when the principal canal was closed for maintenance between October 15 and November 1 each year. As provided by the order of Fernández de Castro, the haciendas outside the city made an annual payment for the maintenance of the system (in the late eighteenth century the Indian community escaped payment[108]) but the money was not always spent for its intended purpose. The Marqués del Villar de Aguila accused the corregidor of keeping the money for his personal use and attempted unsuccessfully to conduct an audit.[109] In 1744, the city obtained permission from the audiencia to levy a general assessment to repair the municipal dam.[110]

Outside the environs of the city the patterns of water use were simpler and less well-documented. A brief sketch will serve to put the city within a broader regional context. At the beginning of the seventeenth century, the pueblo San Francisco possessed a dam and irrigation system on the River Pueblito near its juncture with the River Querétaro. In 1606, the pueblo leased a large share of this water to the local encomendero, Pedro de Quesada. The conveyance recites that the pueblo had recently received a grant of water rights, but no record of the grant appears in subsequent litigation. The water was later reckoned in terms of thirty days, of which twenty days were rented to Spaniards. In 1713, the pueblo made the mistake of terminating its lease with Joseph de Urtiaga, alférez real of Querétaro, and renting the water to another Spaniard. Urtiaga challenged the pueblo's title to the water and secured a grant from the audiencia of twenty days of water.[111]

One finds other evidence of early irrigation in the valley between Querétaro and Apaseo. Four land grants in 1577 refer to irrigation.[112] The Hacienda Tlacote close to

---

[104]In addition to sources cited above, see AGN-H, Vol. 439, Exp. 19, f. 9.

[105]AGN-A, Vol. 138, Cuad. 1, f. 101, Cuad 8, f. 180; Vol. 147, Cuad. 1, f. 125-7, Cuad. 4, f. 1, Cuad. 5, f. 68; Vol. 225, Exp 2, f. 2.

[106]AGN-A, Vol. 225, Exp. 2; AGN-H, Vol. 439, Exp. 19, f. 9v.

[107]Primeras Ordenanzas, passim; AGN-A, Vol. 225, Exp. 2.

[108]AGN-A, Vol. 138, Cuad. 1, 3, and 8; Vol. 225, Exp. 2.

[109]AGN-M, Vol. 72, f. 75, 79v.

[110]AGN-M, Vol. 75, f. 18.

[111]MNA.AM-Qro, roll 5 (Nicolás de Robles, January 7, 1606), roll 51 (Diego de la Parra, June 21, 1712); AGN-M, Vol. 70, f. 14v. For identification of Quesada, see AGN-M, Vol. 22, f. 17; AGN-T, Vol. 307, Exp. 2, f. 73.

[112]AGN-M, Vol. 10, f. 213v.

Apaseo was partly irrigated in 1643, and the Haciendas Castillo and Balvaneda near the junction of the Rivers Querétaro and Pueblito were reputed in the later eighteenth century to have been irrigated from "very remote times."[113] But there was unquestionably an intensification of irrigation in the eighteenth century. The Hacienda Banegras begins to appear on the municipal list of water assessments in Querétaro.[114] The Hacienda La Punta near Apaseo evidently acquired an irrigation system, and several dams were built, most notably a new dam of the Hacienda Castillo that was constructed in 1787 and enlarged in 1797.[115]

At the close of the colonial era, the haciendas along the valley between Querétaro and Apaseo (mostly within the jurisdiction of Celaya) characteristically grew an irrigated winter wheat crop as well as maize, beans, chile, and other summer crops. The water was taken from a series of small dams along the lower reaches of the River Pueblito and the River Querétaro. The haciendas did not possess formal grants of water rights, and there was no system of irrigation turns or regulation of headgates. Irrigation led to little competition over the water. Until a dispute broke out around 1798 between the Hacienda Castillo and the Hacienda La Punta, the hacendados on the river below the junction with the River Querétaro had been content to take the water remaining in the stream after the irrigation of upstream haciendas. The reason is clear: the haciendas all irrigated from storage basins. During the summer flood of the River Pueblito, the water usually was abundant, and all the haciendas could fill their basins without prejudicing the others. During the dry season, the River Querétaro was consumed in the environs of the city, and the river of the Pueblito dried up, effectively precluding the need for allocation.[116]

Irrigation in the valley was augmented by arroyo dams at the base of the hills. There are early seventeenth century records of such dams in Apaseo El Alto and on the Hacienda Jurica near Querétaro. An Indian cacique of Querétaro secured a grant of water rights to the arroyo Juriquilla in 1621, and the hacienda at the base of the arroyo had four fields (tablas) of wheat one hundred years later.[117] The most interesting innovation of the eighteenth century was the dam of the Hacienda Espejo, which made extraordinarily effective use of the water from the Arroyo Mandujano.[118] It will be studied in detail in the chapter on technology.

In the hilly stock-raising country occupying most of the alcald*í*a mayor of Querétaro (and part of the eastern territory of Celaya), water had a very different significance from that in the Bajío proper. All the haciendas had to find means of surviving the long dry season, but the most vital necessity was not irrigation but water to sustain livestock and supply domestic needs. The landscape was dotted with small dams that formed reservoirs or channeled water to ponds (jagüeyes). Some of the dams provided a supplementary source of water for maize milpas, but only a few of the largest dams and

---

[113]AGN-T, Vol. 1353, Exp. 1, Cuad. 1, f. 129v; Vol. 3618, Exp. 1, Cuad. 1, f. 2v.

[114]AGN-A, Vol. 138, Cuad. 8, f. 78v.

[115]The Hacienda La Punta is described as having been put under irrigation more recently than the neighboring Haciendas Tunal and Calera. AGN-T, Vol. 3618, Exp. 1, Cuad. 3, ff. 8v-82.

[116]AGN-T, Vol. 1353, Exp. 1, e.g., ff. 10v, 55v; Vol. 3618, Exp. 1, e.g., Cuad. 2, f. 22v, and Cuad 3, f. 57v.

[117]El Obispado de Michoacán, p. 161; MNA.AH, Fondo Franciscano, Vol. 92, f. 28v; AFC-CSC, leg. "pleitos"; AGN-T, Vol. 440, Exp. 4, f. 64; AGN-M, Vol. 35, f. 124.

[118]AGN-T, Vol. 1353, Exp. 1, ff. 69, 107v.

reservoirs irrigated a winter wheat crop.[119] Springs were greatly prized and generally enclosed with stone fences. Norias fulfilled a vital role, sometimes lifting water from depths of fifty meters. The rural economy depended on perennial sources of water; the hacienda's investments in waterworks often represented an important part of its value. An appraisal of the Hacienda Ojo Zarco, for example, assigned a value of 6,730 pesos to various dams and storage facilities out of a total value of 28,200 pesos (Figure 4.5). But cooperative waterworks were unknown, and individual exploitation of the many small sources of supply rarely led to formal grants or legal adjudication of water rights.[120]

The Valley of San Juan del Río presents parallels both with the rest of the Bajío and with the surrounding stock raising country. The pueblo itself seems to have been founded about the time of Querétaro by Otomí settlers similarly fleeing the Spanish invader. Later in the century, the Indians cultivated not only indigenous crops, but enough wheat to create a need for a water-powered mill. The pueblo received a license to operate the flour mill in 1579, but the population declined and the mill was first abandoned and then sold to a Spaniard.[121] Like the pueblo San Francisco, San Juan del Rio's water soon fell prey to the encomendero, Pedro de Quesada. In 1602, the pueblo conveyed to him the right to one-half the water in the municipal canal (acequia real) (Figure 4.6). Quesada has used Indian labor to build a dam on the River San Juan. The document of conveyance recites that the pueblo gave him the water in gratitude for its supervision of this improvement.[122]

On the other principal tributary system of the Tequisquiapán River, then known as the River of the Estancia Grande, Alonso Pérez de Bocanegra, a descendant of the encomendero of Acámbaro, established a modest wheat hacienda in the 1580s. He was joined by at least two other Spaniards with estates somewhat downstream who competed with him for the unreliable seasonal flow. One Diego Villapaderna claimed to have spent 1,000 pesos employing crews of "16 to 20" Indians for a period of a year to open a ditch to his property; he complained that Bocanegra threatened to divert the stream above the head of the ditch to a flour mill (Figure 4.7). A second neighboring landowner eventually secured a grant of water rights in 1613 to protect his interests.[123]

In the early eighteenth century, the successor of Quesada retained rights to one-half the water of the municipal canal of San Juan del Río, which they used to irrigate three caballerías of land.[124] Part of the pueblo's remaining water was rented to another Spaniard in 1713.[125] On the tributaries of the River Estancia Grande, the Hacienda Llave built a most impressive masonry dam, and several other haciendas irrigated a small

---

[119]AGN-T, Vol. 3401, Exp. 1, Cuad. 1, f. 5v, Cuad. 2, f. 17 (Haciendas San Francisco Xavier de la Barranca y San Lucas); MNA.AH, Gómez de Orozco, leg. 1, ff. 234-243 (LaGriega).

[120]AGN-H, Vol. 72, Exp. 9. Also AGN-T, Vol. 159, Exp. 5, f. 33; Vol. 495, Exp. 1 (both refer to irrigation of milpas); Vol. 1041, Exp. 1, f. 46 (description of noria).

[121]AGN-T, Vol. 307, Exp. 2, ff. 58v, 72; Velázquez, I, p. 39.

[122]AGN-T, Vol. 307, Exp. 2, f. 72.

[123]AGN-T, Vol. 2782, Exp. 3; AGN-M, Vol. 12, f. 110v; Vol. 28, f. 204.

[124]AGN-T, Vol. 307, Exp. 2, e.g., f. 115.

[125]MNA.AM-Qro, roll 51 (Pedro Ballesteros, March 11, 1713).

wheat crop throughout the century.[126] Some irrigated wheat and maize farming also existed from the mid-seventeenth century in the valley of Tequisquiapan to the north and the valley of Amascala near Querétaro.[127]

But these were islands of agriculture in a sea of stock raising. The exploitation of water resources was described in detail in a report that an insightful subdelegate assigned to the pueblo submitted to Viceroy Revillagigedo in 1793.[128] The dominant pattern of water-use had little resemblance to that which we have studied in the rest of the Bajío. The peculiar demands of stock raising had produced a landscape of norias, wells, enclosed springs, small dams, and ponds, for the benefit of livestock, homes, and maize milpas. Waterworks here were exclusive possessions of individual haciendas and no doubt can be best studied in the context of the economic history of the hacienda.[129]

---

[126]AGN-H, Vol. 72, Exp. 9 (Michintepeque, Llave, Estancia Grande); AGN-T, Vol. 272, Exp. 1, ff. 3, 78, 154 (Lira, Zapatilla, Escolastica); Vol. 465, Exp. 1, f. 15. The Hacienda Llave was clearly unirrigated in 1643 (MNA.AM-Qro, roll 23, first doc., f. 144-50) and a description of the hacienda around 1700 does not mentioned irrigation. MNA.AM-Qro, roll 27, Libro de Becerro, No. 1, Item 58.

[127]AGN-M, Vol. 81, f. 128; AGN, Industria y Comercia, Vol. 14, Exp. 2, f. 1; MNA.AM-Qro, roll 23, first doc., f. 165, AGN-T, Vol. 2648, Exp. 3.

[128]But see AGN-M, Vol. 45, f. 266; AGN-T, Vol. 533, Exp. 2 (water needed to water livestock).

[129]AGN-H, Vol. 72, Exp. 9. Also AGN-T, Vol. 673, Exp. 4, f. 18-20.

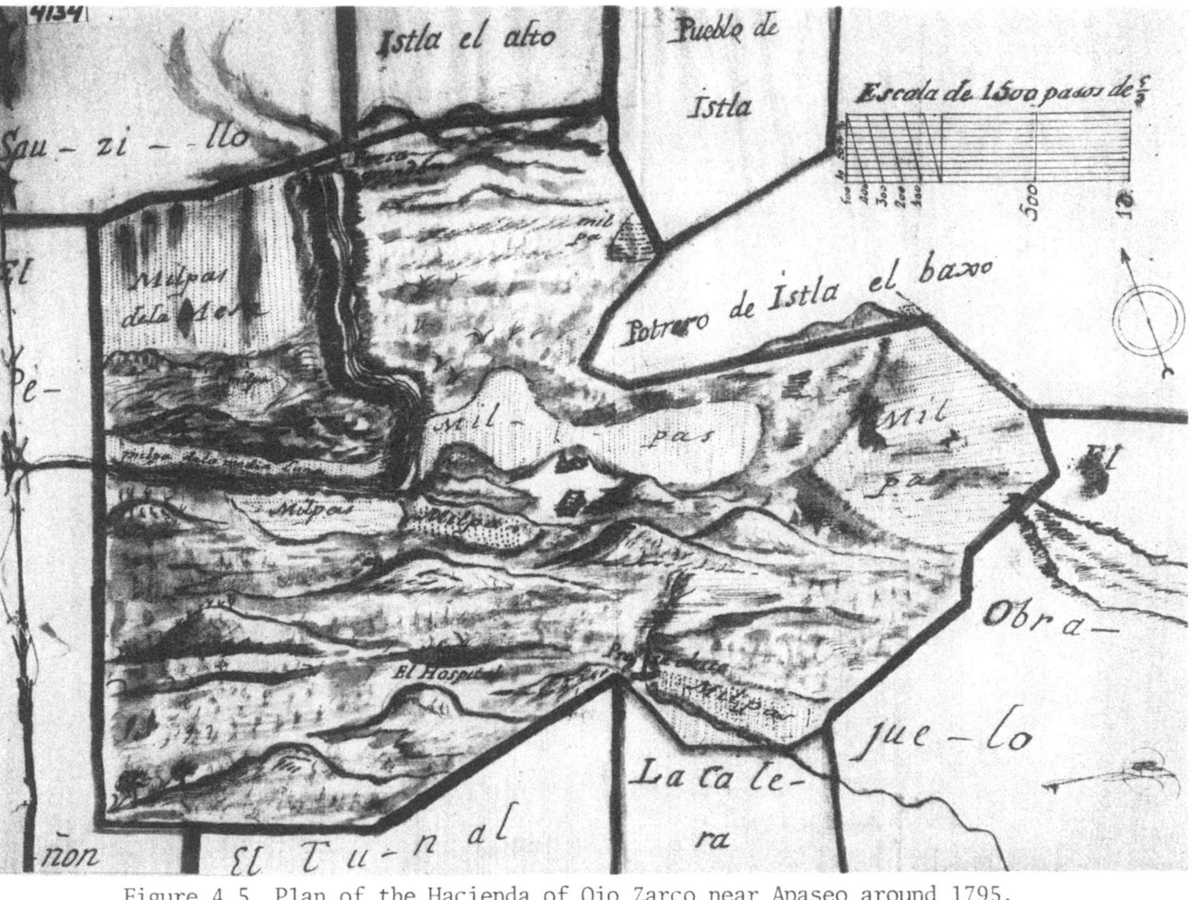

Figure 4.5 Plan of the Hacienda of Ojo Zarco near Apaseo around 1795.
(Source: AGN-T, Vol. 1298, exp. s/n)

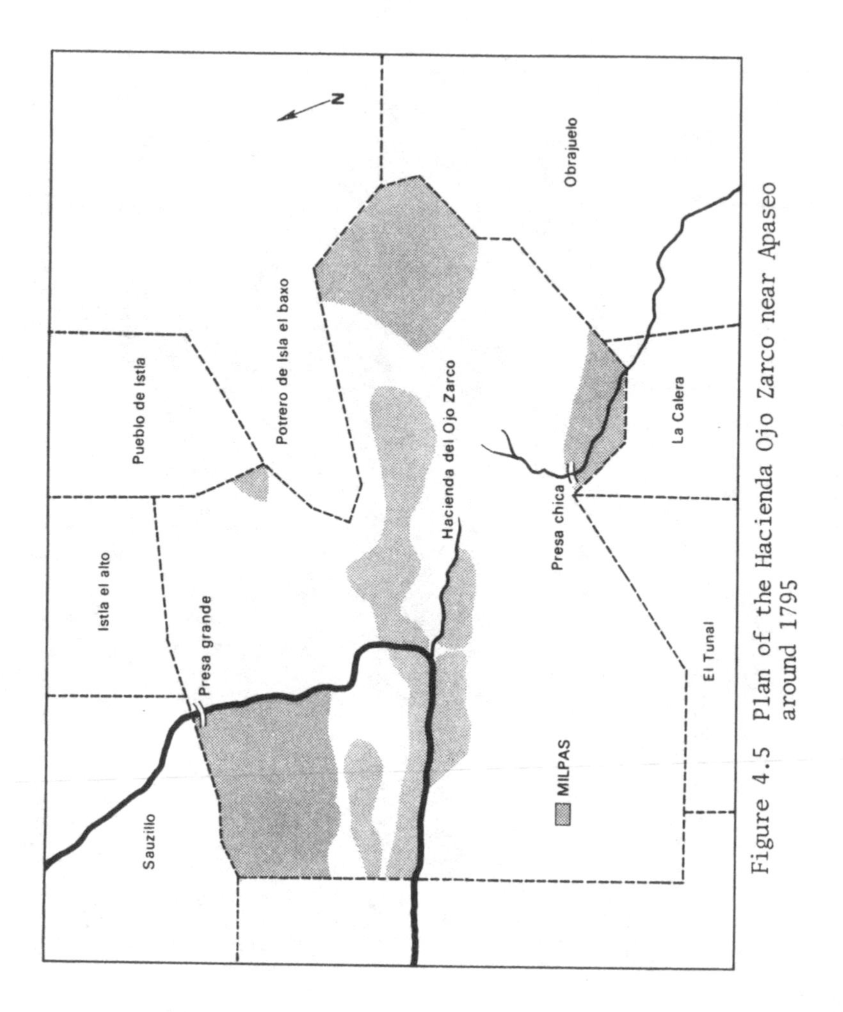

Figure 4.5 Plan of the Hacienda Ojo Zarco near Apaseo around 1795

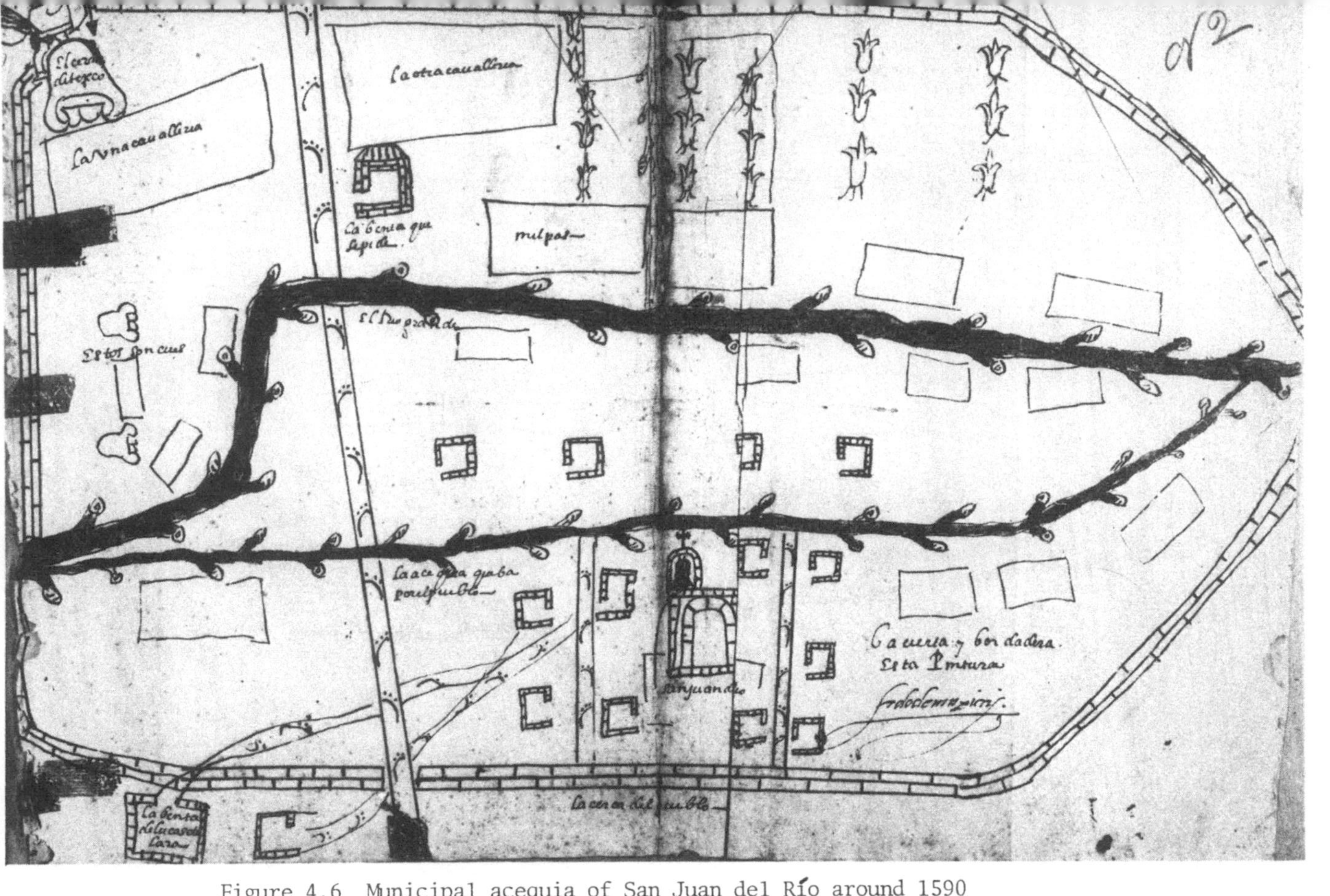

Figure 4.6 Municipal acequia of San Juan del Río around 1590
(Source: AGN-T, Vol. 2782, exp.4)

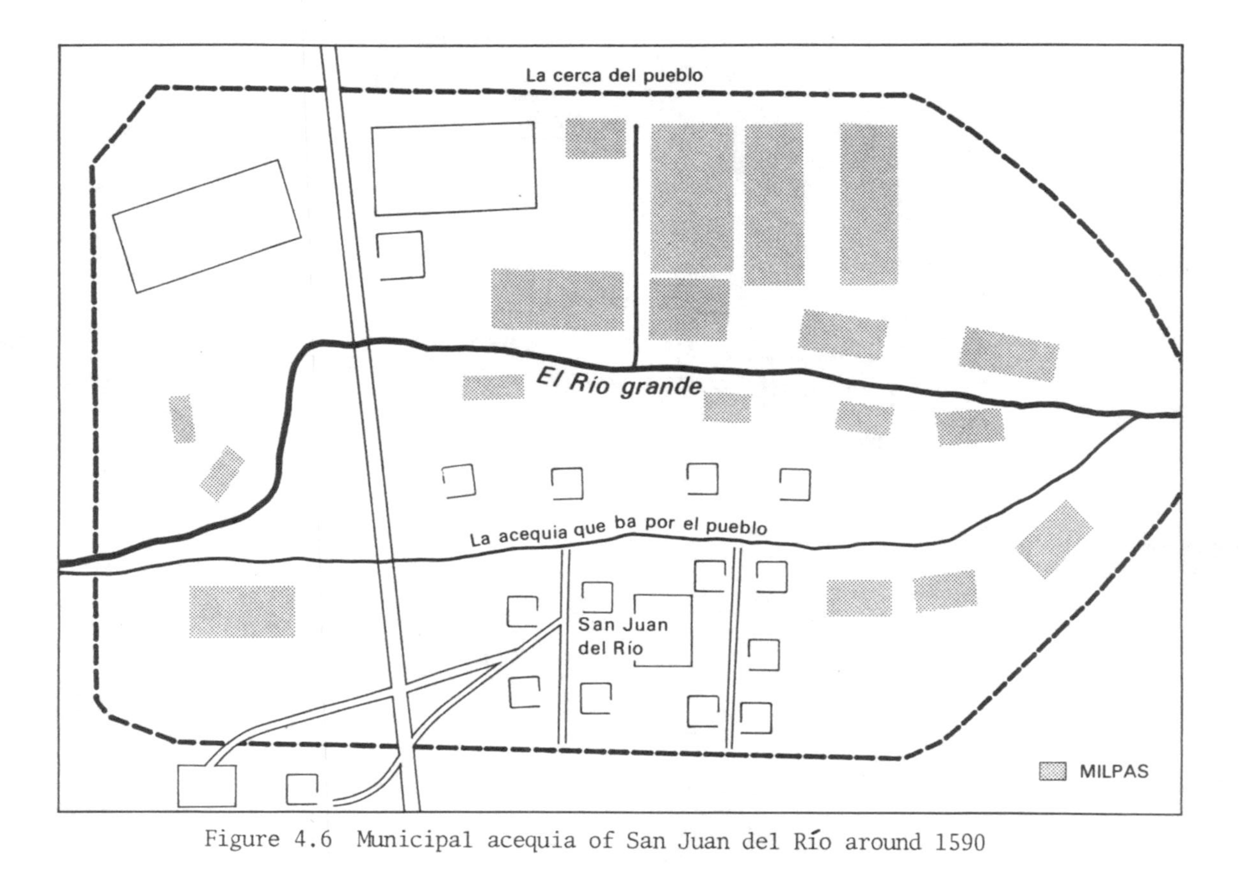

Figure 4.6 Municipal acequia of San Juan del Río around 1590

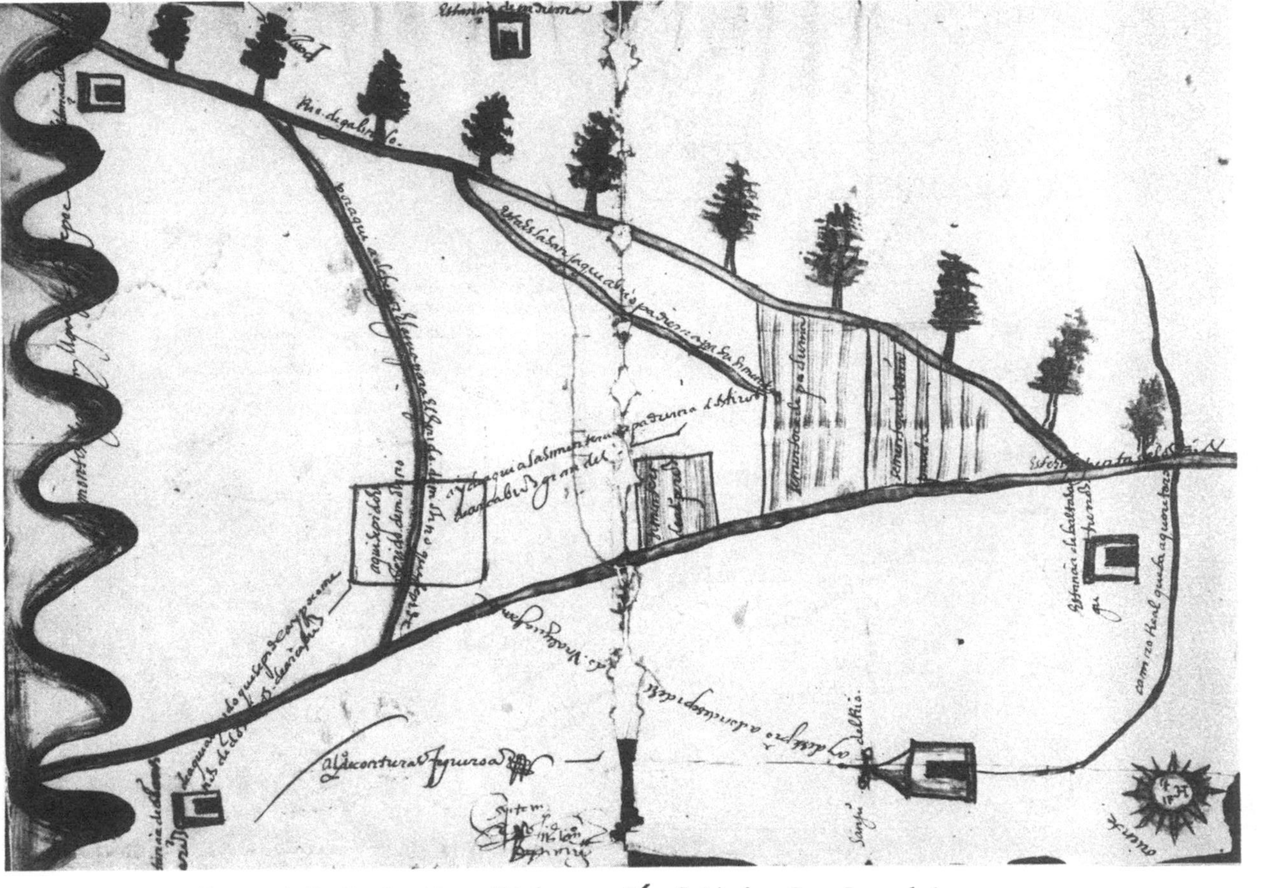

Figure 4.7 Irrigation ditches on Río Galindo, San Juan del Río, in 1584 (Source: AGN-T, Vol. 2782, exp.3)

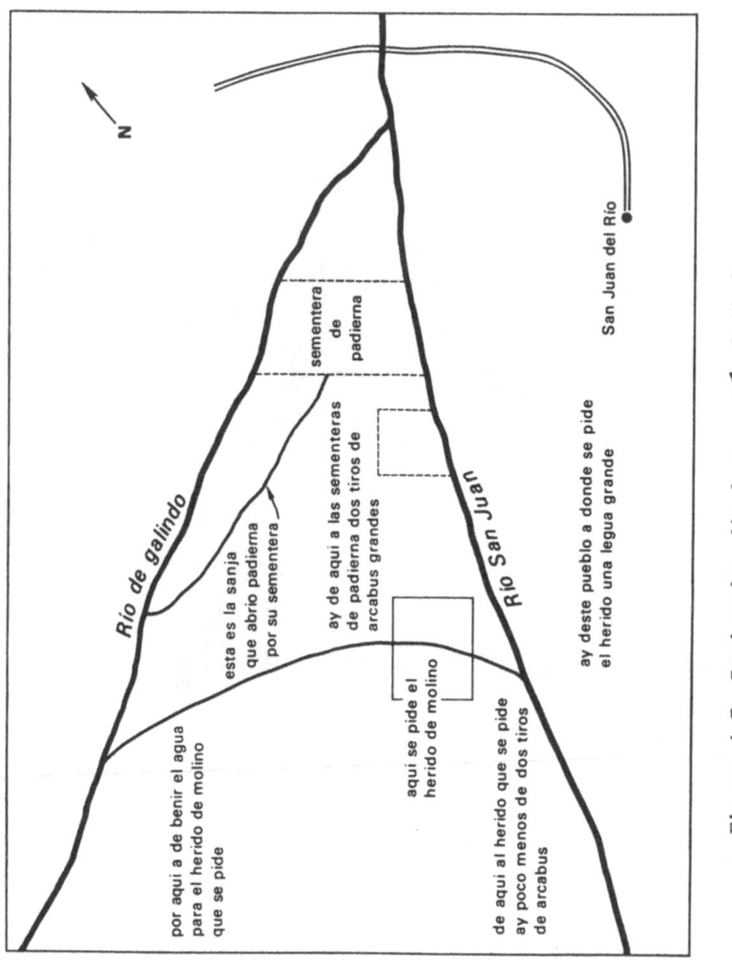

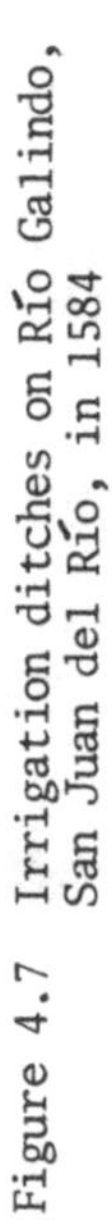
Figure 4.7 Irrigation ditches on Río Galindo, San Juan del Río, in 1584

5
Irrigated Agriculture

Clear evidence that irrigation could, under favorable circumstances, multiply the value of benefited land is provided by eighteenth century appraisals that separately valued irrigated and unirrigated agricultural land on the same haciendas. Such an appraisal of the Hacienda Santa Rita on rich alluvial land north of Salvatierra gave a value of 400 pesos per caballería for unirrigated land, and 1,000 pesos per caballería for irrigated land; the appraiser commented that the value of the irrigated land would have been 2,500 pesos per caballería but for defects in the irrigation system.[1] In the Valle de Santiago, two similar appraisals put values of 300 and 500 pesos per caballería on unirrigated land and a value of 1,500 pesos per caballería on irrigated land.[2] In the region of Celaya, the best comparison is afforded by three evaluations in 1739 of haciendas of comparable land, one lying within the irrigation system and the other two lying just outside. The irrigated hacienda was valued at 2,400 pesos per caballería, and the unirrigated haciendas were valued at 300 pesos per caballería.[3] In Querétaro, the water rights of the Hacienda Doña Melchora were separately appraised at 10,000 pesos. Since the irrigation system served only four caballerías, it contributed a value of 2,500 pesos per caballería.[4] Similarly, the dam on the Hacienda San Lucas was appraised on the basis of the water it provided; the appraiser judged this value to be 2,000 pesos for each of the six caballerías that the dam could irrigate.[5] Because of its proximity to the city, the land of Doña Melchora was

---

[1]AGN-T, Vol. 787, Exp. 1, f. 30 (1755).

[2]ACM, leg. 831, Exp. 450 (San José de Jaral and Mesquite Grande, 1759); ACM, leg. 830, Exp. 425 (Guadalupe and San Joseph, 1752).

[3]AFC-CC, L-4, la pte, Cuad. 4, ff. 19, 38 (Haciendas Santa María, Tenería, and Muñiz). Each of the appraisals of land excludes barns, granaries, and other buildings, but an ambiguous phrase in the appraisal of the irrigated Hacienda Santa María ("con el mueble que contiene la escriptura, sanja y todo lo que le pertenece") suggests that some personal property may have been included in the valuation of its land; such personal property was clearly excluded from valuation of land in the other two cases.

[4]MNA.AM-Qro, roll 51 (Diego Antonio de la Parra, June 26, 1713).

[5]AGN-T, Vol. 3401, Exp. 1, Cuad. 2, f. 22 (c. 1770).

valuable apart from its water rights, but the sort of land that was irrigated on the Hacienda San Lucas had a value of only 150 to 300 pesos per caballería without irrigation.[6]

These examples, compatible with other evidence,[7] show that in the eighteenth century an effective irrigation system could increase the value of land in the Bajío as much as seven or eight times. The high value placed on irrigated land comes as a surprise when one realizes that the climate of the Bajío is only mildly arid and is quite suitable for the unirrigated production of the principal staple, maize. Why good irrigated land was so prized is the first question that we will address in this chapter. It calls for consideration of climatic conditions, the ecology of maize and wheat cultivation, the documentary record of irrigation, and the economics of wheat production. We may then consider the probable extent of irrigation and the connection between irrigation and the level of agricultural productivity.

The Bajío receives substantial and relatively reliable rainfall. The average annual rainfall, according to contemporary records, ranges from 570 mm in Celaya to 730 mm in Salvatierra.[8] The region does suffer from occasional droughts; between 1660 and 1800, widespread hardship resulting from inadequate rainfall is reported for nine years, and certain localities were affected by drought in other years.[9] But the normal year-to-year variations in rainfall are much less than in drier northern Mexico and clearly fall within the limits favorable to the development of rain-fed agriculture. Using the method of C.C. Wallen, the Secretaría de Recursos Hidráulicos reports a relative interannual variability of 24.6 percent for Querétaro, 25.5 percent for Celaya, and 27.2 percent for Salvatierra. Few regions in Mexico have a significantly more stable pattern of precipitation.[10]

As in the rest of the Mesa Central, there is a marked seasonal variation in rainfall. The summer is a season of afternoon thunderstorms, caused by local convection currents, abetted by the rising air of the easterly tradewinds. Most localities experience more than seventy thunderstorms during this season. About eighty percent of the rainfall falls in the four-month period of June through September. In the winter during the period of low sun, the descending air of the subtropical calms generally retards precipitation, although the occasional intrusion of cold air from the north may be accompanied by stormy

---

[6]The appraisal of the four caballerías of land on Doña Melchora at 5,600 pesos probably included undisclosed improvements. The agricultural land of San Lucas was valued at forty-five pesos per caballería, but included hillside land of little value. Bottom land in San Juan del Río, roughly comparable to the land on San Lucas, was appraised at 150 to 300 pesos per caballería in 1727. AGN-T, Vol. 465, Exp. 1, f. 14. Also, AGN-T, Vol. 275, Exp. 1, Cuad. 3, f. 16.

[7]Good irrigated land on the Haciendas La Bolsa and San Rafael in the Valle de Santiago was appraised at 2,000 pesos per caballería in 1762; land in the eastern border of the valley with a less reliable water source and poorer soil was appraised at 1,000 pesos per caballería. ACM, leg. 837, Exp. 495, f. 4. But Brading's tabulation of land values around León shows much narrower variation in the value of irrigated and unirrigated land. Brading (1978), pp. 83-85. A few other references provide conflicting evidence. Morin, pp. 245, 253; Tovar Pinzón, p. 164.

[8]Mexico, SRH (1976), p. 139.

[9]Mexico, SRH (1980), Appendix 3.

[10]Mexico, SRH (1976), p. 114. Using different data, Wallen himself computed an even lower coefficient of variability. Wallen, p. 148.

weather.[11] The Valle de Santiago experienced a typical pattern for the period 1947 to 1976[12] as depicted in Figure 5.1.

It is the high evapotranspiration rate that gives the region an arid character. Statistics published by the Secretaría de Recursos Hidráulicos show annual rates of potential evaporation ranging from 2,945.7 mm in León to 1,632.3 mm in San Juan del Río. The data are based on observed evaporation of water from shallow pans of the sort used by the U.S. Weather Service.[13] These high values reflect the low relative humidity during the dry season and the relatively high temperatures and low incidence of cloudiness throughout much of the year.[14] It may be expected that the actual evapotranspiration, although difficult to measure, corresponds to the high potential evaporation. The moisture received in the first thundershowers in April and May is soon lost to the atmosphere. Throughout the rainy season, the frequent occurrence of brief thundershowers on warm and sunny afternoons tends to maximize evaporation.[15]

The seasonal pattern of evapotranspiration is, of course, affected by variations in temperature. The hottest weather is usually experienced in spring; the summer rains bring some slight relief, and the region's northern latitude causes moderate cooling in winter, accentuated by invasions of polar air. We may use average monthly temperatures (°F) for the Valle de Santiago (1947-76), shown on Figure 5.1, as an example.[16] The high spring temperatures, combined with cloudless days and low relative humidity, result in a period of relatively intense water stress. A study of Chapingo, which has a similar climate, showed that in April, 1973 there was potential evaporation of 216 mm and precipitation of only 30 mm. The summer rains bring some cloudiness and higher relative humidity, but precipitation exceeds potential evaporation for only brief periods. In July and August, 1973, Chapingo had potential evaporation of 135 mm and 116 mm and precipitation of 130 mm and 197 mm.[17]

Despite the mild winter weather, the intrusions of polar air, often coinciding with low relative humidity, regularly cause brief nocturnal frosts. Most districts in the central Bajío reportedly receive an average of more than ten frosts a year in the months of November through February.[18] Early frosts are rare but can have a devastating effect on the ripening maize plant. The incidence of frost precludes the cultivation of maize and certain other crops in the winter but does not damage the wheat crop. Although the leaves of the young wheat plant may be injured somewhat, the shoot continues to grow and, with the aid of favorable spring weather, will yield a good harvest.[19]

---

[11]Vivó Escoto, pp. 187-215.

[12]Mexico, SRH (1979), Table No. 2.4.1.3.

[13]Mexico, SRH (1976), p. 137; Mexico, SRH (1979), Table No. 2.4.3.1; Martínez Luna, pp. 25-78.

[14]See Thornthwaite, pp. 55-94; Penman, pp. 43-50.

[15]See Mexico, SRH (1976), Section 10.03.

[16]Piña Davila, p. 79.

[17]Jauregui, pp. 11-25.

[18]Piña Davila, p. 82; Mas Sinta, p. 19.

[19]Martín, p. 189.

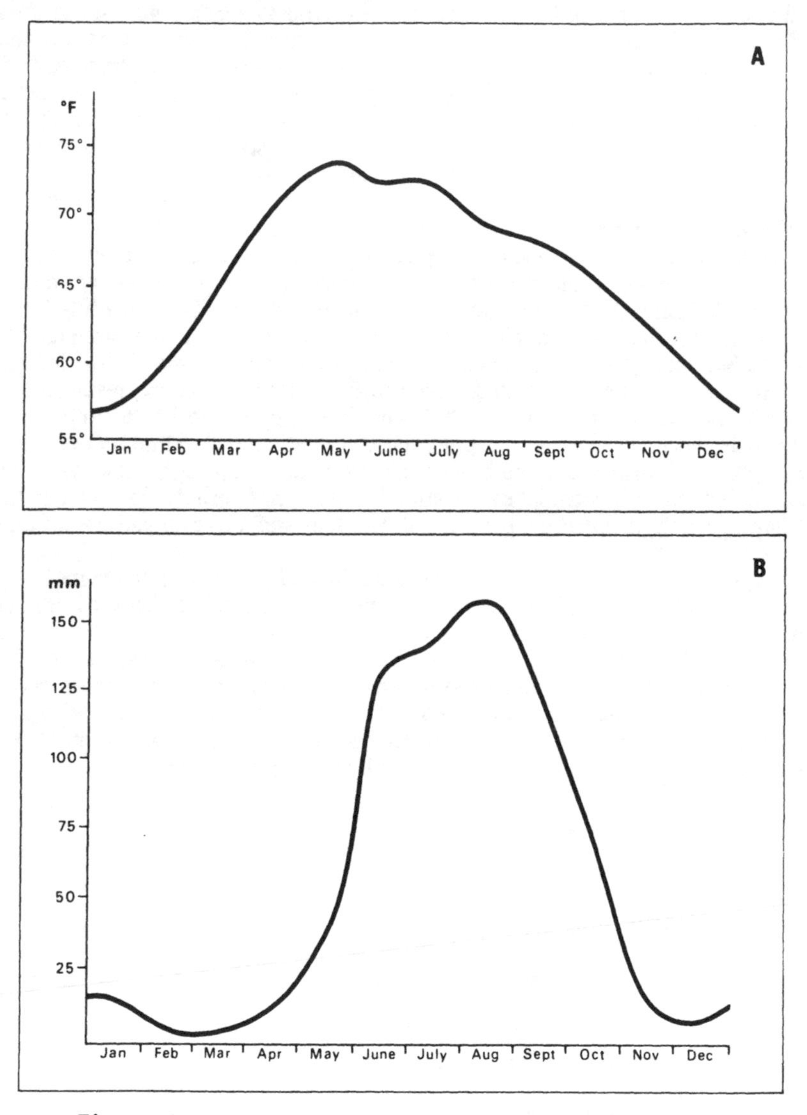

Figure 5.1 Precipitation (B) and temperature (A), Valle de Santiago

The climatic regime of the Bajío is highly appropriate for a warm-season maize crop. In the colonial period, the crop was commonly sown in March or April. The early spring rains promoted germination, and the more abundant rainfall, plentiful sunshine, and warmth of the summer months stimulated growth. The crop could reach maturity from August to October, depending on the time of sowing, the commencement of the rains, and the variety of seed. It stored well in the field and harvesting was commonly postponed to December, January, or even later.[20] Modern varieties of maize grown in highland Mexico mature in three or four months, but the colonial consumers sometimes favored slow maturing varieties. No doubt with some exaggeration, Alzate y Ramírez reported that the Chalco variety popular in Mexico City required six months to mature. But other varieties that matured in as little as three months were also grown and exposed the cultivator to fewer climatic risks.[21]

The summer climate presented obstacles, however, to the cultivation of wheat. The principal varieties imported from Spain evidently could flourish only in an unusually long rainy season and were vulnerable to complete loss in a dry year. In 1582, the corregido of Querétaro explained,

> Little wheat is harvested because there is little irrigation and the rains are poor although for maize they are enough because they continue for three months, and most of the rain is from St. John's day to the beginning of October.[22]

The high temperatures in spring and summer may similarly have been unfavorable to the wheat crop. Since wheat was generally grown in the cool winter season in Spain, it is plausible to suppose that at least some Spanish varieties of wheat were poorly adapted to the warm summer weather, although proof on this point is lacking.[23] The summer thundershowers and occasional hailstorms could also cause lodging of the maturing wheat crop.[24] But most serious was the risk of fungal disease, especially wheat rust, during the hot and humid weather of the summer months. The folk terms, *el chahuistle* and *la roya*, were commonly used to describe wheat rust, although one can find instances in which the former term was applied to insect or worm infestations and to diseases of beans and other

---

[20]*Gazetas de México* II, p. 186; Swan, p. 134; AGN-T, Vol. 550, Exp. 1, ff. 202-30.

[21]Alzate y Ramírez, II, pp. 233-34; Gibson, pp. 307-308; Wellhausen, passim. Humboldt reported a race of maize that matured in two months after sowing. Humboldt, III, Book 4, Chap. 9, p. 60.

[22]Velázquez, I, p. 44. Compare: Prem, p. 53. Prem believes that wheat was planted in the early spring in this high valley so as to avoid the effect of autumn frost.

[23]There have been no studies of the varieties of wheat grown in colonial Mexico, but see Swan, p. 143. Experiments with modern varieties show that the yield is generally reduced in thermal environments over 25°C. Asana (1965), passim; Wattal, passim. But some varieties are known to do best at temperatures of 20°C or lower. Canvin, passim. Excess temperatures are most damaging when the plant nears maturity and yellowing of the leaves may occur. Despite the fact that wheat is commonly grown in regions with a cool, moist spring, experiments show that high temperatures stimulate early growth. Asana, passim; Friend, passim. Rapid early growth may, however, weaken cell structure and make the plant more susceptible to lodging.

[24]See Kakde, passim.

crops.[25] Mota y Escobar, speaking of Guadalajara in 1605, wrote that wheat was not grown in the summer because "unirrigated wheat (*trigo de temporal*) doesn't turn out well but is dark and infested with rust (*prieto y añublado*)."[26] Galván Rivera noted in 1842 that in Mexico unirrigated wheat was not sown on account of the "chuistle."[27] Alzate y Ramírez attempted to discover the cause of wheat rust by examining infected plants under a microscope and observed: "The more humid it is the more abundant is the chahuistle and for this to occur the weather must be warm."[28] Humboldt similarly asserted that the disease which the Mexicans call chahuistle

> often destroys the most abundant crops of grain when the spring and the beginning of summer have been very warm and storms occur frequently.[29]

The cereal rusts are fungal diseases found wherever wheat is cultivated. The three principal categories of rust--stem rust, leaf rust, and striped rust--all exist in modern Mexico, but agronomists associate the traditional names of el chahuistle and la roya chiefly with stem rust (*Puccinia graminis tritici*).[30] No other wheat disease has the same potential for sudden destructive epidemics. It was viewed as a scourge of God in many passages of the Old Testament.[31] The spore germinates on the epidermis of the plant, extends germ tubes into stomata, and expands in pustules under the surface of the stem. Within seven to ten days, the epidermis breaks and new spores are released in great abundance into the atmosphere. The infection causes lodging due to weakened stems, poor seed set, and shriveled grain. Grain harvests are sharply reduced in quantity and are of poor quality.

The spread of rust is greatly affected by weather conditions. Loeggering describes the optimal conditions for the spread of stem rust.

> Free water must be available for six to ten hours for the spores to germinate on the plant surfaces and for formation of oppressoria but is not necessary for penetration. Germination and infection may occur over a wide range of environmental conditions. Germination proceeds best in the dark at temperatures of 60 to 70°F. Penetration

---

[25]Bancroft, III, p. 229; Swan, pp. 169-77.

[26]Mota y Escobar, p. 27. Arregui may have also described stem rust, but his language is obscure ("Si tres dias llueve arreo todos los trigos questan de sacon nazen en la mismi espiga como si estuvieron dabajo de la tierra y esto mas quanto las tierras tocan en calientes"). Arregui, pp. 18-19.

[27]Galván Rivera, p. 121.

[28]Alzate y Ramírez, IV, p. 51. Also Sáenz, *Libro de las Ordenanzas* (mss), f. 65v.

[29]Humboldt, III, Book 4, Chap. 9, p. 107. He added, "On croit communément que cette maladie du grain est causée par des petits insects qui remplissent l'interieur du chaume, et qui empechent le suc nouricieur de monte jusqu'á l'épi."

[30]Ramon Fernández y Fernández (personal communication). Also Mexico, Secretaría de Agricultura y Ganadería, passim.

[31]Chester, p. 313. Also Wiese, p. 39 et seq.

> proceeds best in the light at temperatures from 75 to 85°F. Cool nights with free water in the form of dew or rain followed by sun and rising temperatures are ideal for the establishment of the fungus. After infection has taken place, disease development proceeds best when temperatures are in the 70 to 85°F range and when the wheat plants have plenty of available water. Prolonged periods of drought or cool, wet, cloudy weather are not conducive to the epidemic development and spread of stem rust of wheat.[32]

The rainy summer weather of the Mesa Central of Mexico could hardly be more favorable to the spread of the disease, but in the winter the extended periods of drought prevent germination of the fungus. When winter rains do occur as a result of the penetration of polar air fronts to central Mexico, the temperatures generally descend to levels unfavorable to the fungus. A winter wheat crop is vulnerable only to unseasonably warm and damp weather. In the tithe records one finds a number of years in which a small wheat harvest was described as being of poor quality.[33] Most likely, these were years of rust infestation.

One does not often find mention of unirrigated summer wheat (trigo de temporal).[34] The most common time of sowing this wheat was probably in the spring. The haciendas of San Juan and San Lorenzo in Salvatierra were known to plant some wheat in March in addition to their principal wheat crop grown in the winter months.[35] But wheat was sometimes also sown in the late summer or autumn with the hope that winter rains would bring it to maturity. Such seedings, known as trigo aventurero, were made along the banks of the Laja and Lerma in fields where flood water provided the equivalent of a first irrigation.[36] If the unirrigated wheat came to maturity, it might still be less remunerative than the irrigated crop. Humboldt observed that the quality of unirrigated wheat was inferior, and he reported that a hacienda near Celaya experienced yields of forty to fifty to one for wheat sown on irrigated land, but only fifteen to twenty to one on unirrigated land.[37] In the seventeenth century, the unirrigated wheat crop sold for ten to thirty percent below the price of the irrigated crop.[38] Tithe records help us to estimate the relative importance of unirrigated wheat production. A few seventeenth

---

[32]Loeggering, p. 314.

[33]ACM, leg. 835 (1735), leg. 838 (1700, 1706), leg. 847 (1746), leg. 860 (1711, 1718).

[34]*El Obispado de Michoacán*, p. 162 ("algo de maiz y trigo de temporal"); Velázquez, I, p. 44 ("poco trigo"). Compare: Mota y Escobar, p. 27, Gemelli Carreri, p. 33; Ewald, p. 19.

[35]Swan (1977), p. 145.

[36]ACM, leg. 834, exp 462, f. 14; MNA.AH Map Collection, leg. 3698, doc. 8 (legend of map). Also AGN-M, Vol. 10, f. 3 (floodplain farming, probably of wheat); AGI, Indiferente, leg. 107, tomo 1, f. 386 ("un poco de trigo aventurero," Irapuato). Arregui describes wheat "de medio riego" planted in August or September and brought to maturity with the aid of irrigation, but I have not found any other documentary mention of this practice. Arregui, p. 19.

[37]Humboldt, III, Book 4, Chap. 9, pp. 82, 105.

[38]ACM, leg. 861 (1673), 838 (1683, 1685, 1690), leg. 860 (1690).

century tithe records expressly distinguish between the irrigated and unirrigated wheat crops; the unirrigated crop varied from less than one percent to as much as twenty-eight percent of the total.[39] The eighteenth century records of Celaya show that a number of haciendas beyond the apparent reach of irrigation produced small quantities of wheat in some years but none in others. We may surmise that this portion of the wheat crop was unirrigated; it was generally below ten percent of the wheat production as a whole.[40] The marginal unirrigated wheat production may help explain the description of apparently unirrigated land as being of *pan sembrar*[41] or *pan llevar*,[42] but the latter phrase was

---

[39]ACM, leg. 860 (1690), leg. 861 (1663), leg. 838 (1676). Some tithe records refer to unirrigated wheat as *trigo candial*, possibly a reference to a distinct variety of wheat.

[40]ACM, leg. 841; MNA.AM-ACM, roll 775532.

[41]The unirrigated Hacienda Almanza was described as consisting of "tierras de pan sembrar." ACM, leg. 831, Exp. 445. In other cases the term was applied to land with, at best, limited irrigation. AGN-T, Vol. 824, Exp. 2, f. 24; Vol. 3687, Exp. 8 (map).

[42]E.g., AGN-T, Vol. 548, Exp. 2, f. 156; Vol. 637, Exp. 3, Cuad. 2, f. 60; Vol. 1061, Exp. 4, ff. 39, 102; AGN-H, Vol. 72, Exp. 9, f. 46v.

highly ambiguous; although it was commonly applied to wheat land, it was also used to describe good quality bottom land devoted to other crops.[43]

The major wheat crop was grown under irrigation in the winter months. The crop was commonly sown around the first of November and harvested in early May, but one finds records of sowing as early as September 19 and as late as December 10 and records of harvesting from April 20 to June 5.[44] Humboldt described the characteristic method of irrigation.

> In the farms (haciendas de trigo) in which the system of irrigation is well established, in those of Silao and Irapuato near León, for example, the wheat is watered twice: first, when the young plant springs up in the month of January; and the second time in the beginning of March, when the ear is on the point of developing itself. Sometimes the fields are also inundated before sowing. One observes that in allowing the water to remain several weeks in the field the soil is so saturated with humidity that the wheat resists

---

[43]I have found four interpretations of the colonial expression "tierras de pan llevar." First, a correspondent of the Sociedad Méxicana de Geográfia y Estadística defined the term as meaning arable hillside land in a report on Querétaro in 1848. Raso, p. 20. This definition has been picked up by John Super in his very interesting history of Querétaro. It is, however, quite inconsistent with the actual usage of the term in the colonial Bajío, including the region of Querétaro (e.g., AGN-T, Vol. 307, Exp. 2, f. 102; AGN-H, Vol. 72, Exp. 9, ff. 46v, 48).

Second, Galván Rivera made a rather confusing attempt at definition in 1842 that appears to define tierras de pan llevar as irrigated or irrigable land (Galván Rivera, p. 121). Michael C. Meyer tells me that this definition conforms to the actual usage of the term in northern Mexico, where tierras de pan llevar were typically devoted to irrigated crops. But in the colonial Bajío, the term was applied to land that is explicitly described as being "de temporal"; to land that was irrigated in only very small part; to land that was currently unirrigated and probably not physically irrigable; and to large expanses of land and to tablelands that were plainly not irrigable.

Third, the term is often understood as meaning wheat land. At first impression, it would seem to be no more than a variation of expressions such as *labor de pan*, *tierras de pan sembrar*, and *tierras de pan sembrado*, which were plainly applied to wheat land (e.g., AGN-T, Vol. 1403, Exp. 2, f. 11v (*pan sembrar de riego*); AGN-G, Vol. 2, Exp. 169; AGN-M, Vol. 29, f. 122; C-AD, Fondo 507-2 ("Tierras de pan sembrar...y su acequia"). In the colonial Bajío, tierras de pan llevar were, in fact, commonly devoted to wheat, and sometimes the association between the term and wheat production was so direct as to suggest that it was used in this narrow sense. The use of the term to describe unirrigated land at least suggests the possibility that the land was considered suitable for trigo de temporal or trigo aventurero. But the large expanse of unirrigated land described as pan llevar in the documents cited above was clearly not devoted primarily to wheat production, and the term is sometimes applied to land with only a very minor wheat crop.

I think Basilio Rojas, himself a farmer in the Bajío and an informed student of colonial history, has provided the best definition of the most common usage of the term. He defined these lands as "aquellas de gran fertilidad y de facil cultivo, pero muy principalmente aquellas que pudieren ser regadas artificialmente."

[44]E.g., AGN-T, Vol. 353, Exp. 2, ff. 6v, 15v; Vol. 550, Exp. 1, ff. 201v-230; Vol. 1353, Exp. 1, Cuad. 5; Vol. 674, Cuad. 2, f. 1149v; AGN-M, Vol. 10, f. 19v; Vol. 70, f. 109v; Raso, p. 35; Swan, p. 143.

> more easily long droughts. The seed is sown by scattering it (on seme a la volée), at the moment when the water is allowed to drain from the fields. This method brings to mind the cultivation of wheat in Lower Egypt, and these prolonged inundations diminish at the same time the abundance of parasitic weeds.[45]

A first irrigation in November appears to have been a more common practice than Humboldt suggests. One often finds assertions that three irrigations were required for wheat; where an irrigation system was shared by several haciendas, each hacienda might receive three irrigation turns.[46] The first irrigation was sometimes the prolonged flooding before sowing that Humboldt described,[47] but it might also consist of a brief and systematic watering. The Hacienda San Juan and San Pablo near Querétaro irrigated between November 15 and 20 and the corresponding days of January and March.[48] Irrigation in the Bajío has long been carried out either by flooding narrow strips within fields or by flooding entire fields comprising several hectares. In colonial documents one finds descriptions of rural properties that suggest the use of both these methods of irrigation.[49] The flooding of fields at infrequent intervals obviously minimized the demand of irrigation for labor.

These irrigation practices seem to have been peculiar to the Bajío. The Bajío soils, with their high clay content, had unusually good moisture retention.[50] A manual for administrators of Jesuit haciendas regarded three irrigations as a bare minimum,

---

[45]Humboldt, III, Book 4, Chap. 9, p. 78.

[46]AGN-T, Vol. 1169, Exp. 1, f. 3; Vol. 1175, Exp. 4, f. 2; Vol. 1353, Exp. 1, f. 37; Vol. 2967, Exp. 112, f. 303; Vol. 3618, Exp. 1, Cuad 2, f. 7. But see Ward, I, p. 46 (two irrigations in January and March).

[47]AGN-T, Vol. 353, Exp. 2, f. 6 (irrigation "previas a las siembras"); Fernández y Fernández, p. 430.

[48]AGN-T, Vol. 764, Exp. 3, Cuad. 3, ff. 22v, 28. Similarly, there is a record of the Hacienda Espejo near Querétaro irrigating during the periods November 3 to 22, February 5 to 7, and March 14 to April 2. AGN-T, Vol. 1353, Exp. 1, Cuad. 4, f. 23.

[49]I was assured by Don Javier Fernández de Ceballo, who formerly administered haciendas near Querétaro, that irrigation has traditionally been performed by "inundación" in that region. A private report of the Secretaría de Recursos Hydráulicos in Celaya written in 1922 describes the methods of "submersión" of strips three to ten meters wide and of "entarquinamiento" applied to fields of five to twelve hectares in size. SRH file "Generalidades 'Río Laja.'" The ditch dams and property descriptions on the outskirts of Salvatierra suggest the method of "submersión." E.g., AGN-T, Vol. 988, Exp. 1, Cuad. 3, f. 144. The Hacienda del Molinito near Celaya may have practiced the method of "entarquinamiento." AGN-T, Vol. 2072, Exp. 1, Cuad. 5, f. 21, and illustration. The six cajas shown on the illustration may have been used for irrigation as well as water storage; the hacienda's irrigation system was said to be used to "enlamar" the soil.

[50]For classification of soils according to the FAO method, see Cartas Edafalógicas of the Comisión de Estudios del Territorio Nacional (CETENAL).

suitable only for "tierras frias, pingüez y jugosas y llanos."[51] In the north of Mexico, wheat was irrigated as many as eight times,[52] and furrow irrigation was practiced. I have not found any evidence of furrow irrigation in the Bajío until modern times.[53]

Other secondary crops were sometimes irrigated in association with wheat. One document refers to irrigated barley grown in the winter months.[54] Other sources mention the irrigation of chile in the dry season (*estación de secas*) and in the months preceding the sowing of wheat.[55] Where irrigation systems used storage basins or reservoirs, wheat production was directly associated with the cultivation of garbanzos and lentils. If planted in saturated soil, both should yield a harvest without further irrigation. The crops could thus be planted in storage basins as soon as they were emptied or along the edge of receding reservoirs.[56] Storage basins emptied in March could also be planted with maize, but this practice was likely to conflict with the use of the basins the following year.[57]

The use of irrigation for maize production, though much less important than for wheat, was not uncommon in the colonial Bajío. One finds occasional references to *maiz de riego* on pueblo land or on wheat haciendas.[58] At the base of the hills surrounding the Bajío, the water from small dams and springs were sometimes devoted to maize cultivation.[59] Two haciendas in the Amascala Valley between Querétaro and San Juan del Río possessed irrigation systems that were applied exclusively to the benefit of a maize crop.[60] In a practice known as *medio riego*, water collected in dams and reservoirs during the summer could be released in the early autumn to help bring the maize to maturity.[61] Alternatively, farms enjoying access to a perennial source of water could

---

[51]Chevalier (1950), p. 145.

[52]Basave Kunhardt, p. 204.

[53]Furrow irrigation of wheat was practiced in Chihuahua in 1834 and near Celaya in 1922. Basave Kunhardt, p. 204; private report cited in footnote 49.

[54]AGN-T, Vol. 550, Exp. 1, f. 201; Swan, p. 147.

[55]AGN-T, Vol. 307, Exp. 2, f. 73; Vol. 534, 2a pte, Exp. 1, f 1; Vol. 1113, Exp. 1, Cuad 1, f. 27; Vol. 1175, Exp. 4, f. 62; Vol 1353, Exp. 1, f. 1v; Vol. 2716, Exp. 3, f. 1; Vol. 3618, Exp. 1, Cuad. 3, f. 4v.

[56]Although I have not found any documentary record of this practice, it was universally followed in the early twentieth century and was undoubtedly found in earlier times. I owe this information to personal interviews with four elderly agriculturalists: Luis Alvárez, José Rodríguez, Javier Fernández de Ceballos, and Basilio Rojas.

[57]Javier Fernández de Ceballos (personal communication).

[58]AGN-T, Vol. 630, Exp. 1, f. 56; Vol. 1362, Exp. 1, Cuad. 3, f. 94; MNA.AH, Eulalia Guzmán, leg. 107, doc. 7; AGN-T, Vol. 2716, Exp. 3, f. 1.

[59]AGN-H, Vol. 72, Exp. 9, f. 17; AGN-C, Vol. 1298, Exp. s/n, f. 35 (map of Hacienda Ojo Zarco); AGN-T, Vol. 159, Exp. 5, f. 33; Vol. 495, Exp. 1.

[60]AGN-T, Vol. 2648, Exp. 3, ff. 48, 89.

[61]AGN-H, Vol. 72, Exp. 9, f. 12; AGN-T, Vol. 1353, Exp. 1, Cuad. 4, f. 23. For meaning of term "medio riego," see Arregui, p. 19.

irrigate their maize fields in the late spring and summer.[62] Springtime irrigation was not only a protection against drought but also against autumn frosts. Though infrequent, early frost was a factor to be reckoned with. In 1793, the subdelegate for San Juan del Río wrote that

> If the rains don't arrive in time, the crops are delayed (as occurs in some years) and the frost can catch the tender maize plants and damage them so that they do not produce a crop as occurred throughout this jurisdiction the past year of ninety-three when not a twentieth part of the crop expected at time of sowing was harvested because the rains failed to come in their normal season.[63]

The list of crops that were at least occasionally irrigated could be greatly extended. The irrigated gardens in cities, villages, and rural settlements grew a wide variety of vegetables and fruits. The Valle de Santiago produced irrigated crops of melon, saffron, and sweet potato. Morin sees this production as an example of the speculative cultivation of high-value crops,[64] but a survey of other uses of irrigation only serves to emphasize the central importance of winter wheat cultivation. The early canal systems of the Bajío were all established to grow wheat, and new land put under irrigation in the eighteenth century was ordinarily converted to wheat production.[65] The close association of wheat and irrigation existed around León and Irapuato as well as in the areas we have studied more closely. A map depicting the dams on the Rivers Silao and Guanajuato in 1792 is entitled "Mapa descripción geográfica de los ríos de que toman agua las vasos o presas de la congregación de Irapuato en la Intendencia de Guanajuato para sus siembras de trigo."[66]

Why then was irrigated land so highly prized? It was valued for the opportunity that it afforded for double cropping in only a few intensively cultivated localities. Double cropping was important in the land around Querétaro and in some Indian pueblos,[67] but it seems to have been uncommon elsewhere. We may doubt whether the early settlers in the sixteenth century, enjoying an abundance of land, were interested in double cropping. The early tithe reports of the Valle de Santiago, in fact, show that the irrigated wheat haciendas produced very little maize.[68] In the later colonial period, the agriculturalists

---

[62]AGN-T, Vol. 187, Exp. 2, f. 150v (Apaseo); Vol. 2648, f. 63 (Querétaro); AGN-M, Vol. 38, f. 180 (en tiempo de aguas se riegan).

[63]AGN-H, Vol. 72, Exp. 9, f. 1v. There were destructive frosts in the Bajío in 1661, 1695, 1749, and 1785. Mexico, SRH (1980), Appendix 3.

[64]AGN-A, Vol. 97, Exp. 2, f. 7 (melones y azafrán); Morin, p. 253 (azafrán y camote).

[65]E.g., AGN-T, Vol. 1168, Exp. 3, Cuad. 2, f. 1v; Vol. 1169, Exp. 1, Cuad. 5, f. 113; Vol. 1175, Exp. 4, f. 62v; Vol. 2767, Exp. 3, Cuad. 5, f. 1; AGN-M, Vol. 79, f. 17.

[66]AHML, caja 1606-1609, Exp. 18, caja 1664-1672, Exp. 1; AGN-M, Vol. 60, f. 87; AGN-T, Vol. 1167, Exp. 1, f. 509.

[67]AGN-T, Vol. 187, Exp. 2, f. 150v; Vol. 1353, Exp. 1, Cuad 1, 4, and 5; Vol. 2648, Exp. 2, f. 63. Compare: Lee, p. 650. Lee claims that two wheat crops were grown in the valley of Mexico.

[68]ACM, leg. 860.

continued to lack an incentive to cultivate the land intensively. The colonial farmer spread seed thinly over his fields to achieve a high yield per seed but a low yield per unit of surface. Morin explains, "The abundance of uncultivated land within, as well as outside, the borders of the farms justified this wasteful use of land."[69] The colonial farmers often found it convenient to store maize in the field after the time that wheat was to be sown. On the Hacienda San Francisco Xavier in the Valle de Santiago, the maize was harvested a month after the wheat was sown on separate land.[70] The Haciendas San Juan and San Lorenzo in Salvatierra followed a similar practice.[71] The growing season for slow maturing varieties of maize might occasionally overlap with the wheat season. Moreover, the prospective profits of a maize crop might not justify the purchase of seed. The renter of a wheat hacienda on Apaseo lost his maize crop in 1605 and sowed no maize at all the next year.[72] The tithe reports in the later colonial period show that wheat-growing haciendas generally produced a substantial maize crop, but one sometimes finds records of a large wheat harvest followed by a small harvest of maize.[73] These considerations, together with the absence (in my research) of any mention of double cropping apart from a few islands of intensively cultivated land, support the conclusion that the maize crop and the irrigated wheat crop were ordinarily grown in different fields.

Irrigation had important but limited value as a protection against climatic fluctuations. One finds accounts of irrigated haciendas along the Laja and Lerma that survived drought years more successfully than their neighbors,[74] but the colonial irrigation systems were themselves severely affected by adverse fluctuations in rainfall. In dry years, the principal dam on the Laja failed to provide the farmers of Celaya with a third irrigation;[75] storage basins along the River Querétaro remained empty;[76] the spring water in Apaseo ran dry in March;[77] and even along the Lerma, the region best endowed with water, there were complaints of shortages.[78] Haciendas relying on minor arroyos, such as La Griega in Querétaro, had enough water to irrigate only in normal years.[79] For rich landowners, the best strategy to survive climatic fluctuations was not irrigation, but

---

[69]Morin, p. 242.

[70]AGN-T, Vol. 550, Exp. 1, ff. 201v-230.

[71]Swan, p. 203.

[72]AFC-CSC, leg. "pleitos."

[73]E.g., ACM, leg. 841 (See Haciendas Santa Rita, Plancarte, and Roque in years 1781, 1782, and 1786).

[74]E.g., AGN-T, Vol. 2071, Exp. 1, f. 4 (En los años esteriles como este y '86 todos los que regaran alzaran maiz).

[75]AGN-M, Vol. 39, f. 203v; AGN-T, Vol. 1168, Exp. 3, f. 46v.

[76]AGN-T, Vol. 3618, Exp. 1, Cuad. 3, f. 46v.

[77]AGN-T, Vol. 187, Exp. 2, f. 217v.

[78]AGN-M, Vol. 81, f. 50v; AGN-A, Vol, 97, Exp. 2, f. 10; MNA.AH, Eulalia Guzmán, leg. 107, doc. 7.

[79]MNA.AH, Gómez de Orozco, leg. 1, ff. 234-43.

the construction of grain storage barns, and that was, in fact, a major direction of investment.[80]

The explanation of the value placed on irrigated land can be found principally in the economics of wheat production. Wheat always brought a much better price than maize in the colonial period. In the Valley of Mexico during the early 1570s, maize sold for 4.8 reales per fanega and wheat for 10.1 reales per fanega; this relationship of 1:2 may well have been characteristic of the sixteenth century.[81] Enrique Florescano has undertaken a comprehensive study of the tithe records for the archbishopric of Michoacán that may be expected to provide, among other things, a definitive exposition of price relationships from the mid-seventeenth century to the last decades of the eighteenth century. My own sampling, together with published studies of the tithe records of Dolores Hidalgo and San Miguel Allende, indicate that while there was no standard relationship between the wheat and maize prices, a fanega of wheat often sold for about 1.5 times the price of a fanega of maize in the late seventeenth century and for twice or more the price of maize in the eighteenth century.[82]

Why did wheat sell for a higher price than maize? It seems doubtful that differences in the costs of production often equaled the difference in price. It is true that the yield of wheat was much lower than maize;[83] wheat had to be harvested immediately on maturity and was more difficult to store;[84] and irrigation entailed many expenses. But on the whole, wheat required less labor than maize per unit of production.[85] The price differential between the crops surely exceeded any additional costs incurred by efficient wheat producers. Wheat haciendas were not a uniquely profitable investment; Tovar Pinzón's study of Jesuit haciendas suggests that they ran the gamut of high to low profitability.[86] But if we may assume that the price of wheat, though subject to municipal intervention, tended to have an economic rationale, it perhaps reflected the costs of marginal producers (the farms producing an unirrigated wheat crop or possessing expensive or inefficient irrigation systems). We may easily imagine that wheat haciendas with effective irrigation systems offered the promise of unusually high revenues per unit

---

[80]Florescano (1976), p. 87; *Diario de México* XII, pp. 660-62 (1810).

[81]Borah and Cook, pp. 18-23. Also Mota y Escobar, p. 27.

[82]Hurtado López (1974), pp. 41, 50. Her statistics are apparently based on an equivalency of two fanegas to one carga. I have adjusted the figures for an equivalency of three fanegas to one carga. The colonial measures of volume were subject to considerable regional variations and in the later colonial period were often used as measures of weight as well as volume. The tithe records of Michoacán account for wheat in cargas and maize in fanegas. They clearly reveal that three fanegas of wheat equaled one carga. In the absence of evidence to the contrary, I have assumed that a fanega of maize and a fanega of wheat are equal and that both refer to volume rather than weight. See generally, Florescano (1969), pp. 71-77. On prices and price movements, see also Morin, pp. 188-202, and Galicia, pp. 51-52, 72-74.

[83]Morin, pp. 237-41.

[84]On storage, see Humboldt, III, Book 4, Chap. 9, p. 106.

[85]Swan, p. 144. Swan charts the seasonal labor required for wheat and maize cultivation in his Figures 11 and 12. Also García Miranda, p. 496.

[86]Tovar Pinzón, pp. 184-93.

of land and that this higher economic productivity was reflected in the higher price of the land.

In the eighteenth century, wheat haciendas were undoubtedly valued as a means of agricultural diversification. Through the research of Florescano, we now understand that the landowners' struggle for security lay at the center of the agrarian history of Mexico. He writes:

> Thus one of the obsessions of the hacendados was to obtain a fixed and constant income instead of large profits in one year and losses in another. But since this desired stability was thwarted by the vagaries of weather, the fluctuations of production and the limited demand, the solution of the great landowners was to endow their haciendas with the resources needed to offset the effects of these destabilizing factors. These resources were secured by the territorial expansion of the hacienda which served to bring within its boundaries the greatest variety of land (irrigated, rain-fed, pasture) and natural resources (rivers, springs, woods, quarries).[87]

Wheat was subject to largely different meteorological hazards than maize and most other crops. When the rain-fed crops failed, the wheat crop might do well or vice versa. In the famine years of 1785, for example, the wheat farmers of Celaya produced a bumper crop.[88]

Was wheat production also attractive as a speculation? Morin suggests that imported agricultural products were subject to wide price variations in outlying parts of Michoacán.[89] Wheat was among the crops transported to these relatively distant markets. Alternatively, did wheat producers enjoy a relatively stable market? The tithe records reveal wide local fluctuations in production, but wheat prices may still have been more stable than maize prices. Consider, for example, the cyclical price movements in Dolores Hidalgo in the period 1740-1790:[90]

Amplitude of Cyclical Price Movements of Maize

Cycle	Lowest annual price reales/fanega	Highest annual price reales/fanega	% Relative difference
I	2.5	29	1060
II	3	11	300
III	4	18	350
IV	5	14.5	190
V	8	40	400

---

[87]Florescano (1976), p. 96.

[88]ACM, leg. 841.

[89]Morin, p. 196.

[90]Hurtado López, pp. 40, 49.

Amplitude of Cyclical Price Movements of Wheat

Cycle	Lowest annual price reales/carga	Highest annual price reales/carga	% Relative difference
I	8	68	750
II	28	60	144
III	24	52	116
IV	40	64	60
V	37	72	94
VI	48	116	141

Various factors may have favored relative price stability of wheat. The broad regional markets of wheat may have spread risks affecting price movements, and the fact that wheat was a high-priced alternative to maize, desired by the growing Hispanic population, may have resulted in a comparatively elastic demand. In years of bad harvests, the demand for wheat may have been dampened not only by higher prices but by the poor quality often associated with a small crop. Moreover, the wheat market was free of the destabilizing influence of subsistence producers who affected the market for maize. The small subsistence farmers who produced much of the maize crop sold only their surplus production; in poor years they had no excess to sell, while in good years they flooded the market.

For a better understanding of the economics of wheat production, we must await further analysis; the subject has been little studied. Our discussion of irrigation on wheat haciendas serves only to pose questions for further research.

The Indian pueblos and barrios of the Bajío valued irrigation for somewhat different reasons than the Spanish haciendas. Being reduced to small plots of land, the Indians had every incentive to use them intensively. Irrigation permitted double cropping, increased the yield per unit of land, and assured the success of the maize crop by bringing early germination in dry years.[91] The Indians cultivating the small fields in the Cañada near Querétaro declared that they could not sustain themselves without irrigation.[92] Most allusions to irrigated land in Indian communities refer to huertas, milpas, or *sementeras*, cultivated largely for subsistence, but there are also many references to a wheat crop.[93] The Indians of San Francisco del Rincón, who fought a neighboring hacienda to retain rights to water, explained that

---

[91]E.g., AGN-T, Vol. 2716, Exp. 3, f. 1.

[92]MNA.AM-Qro, roll 23, first doc., f. 277v.

[93]AGN-T, Vol. 294, Exp. 1, f. 20v (Yuririapúndaro); Vol. 307, Exp. 2, f. 72 (San Juan del Río); Vol. 2716, Exp. 3, f. 1 (Etaquaro); CAAM, Book 374 (San Pedro de la Cañada, San Juan del Río, Pueblito de San Francisco); AMS Registro de Tierras 1717 (Taristarán); ACM, leg. 841 (San Miguelito and Neutla in 1783), leg. 836 (Tarandaquaro in 1797), leg. 861 (Urireo, Emerguaro, and barrios of Salvatierra in 1765). In most of these cases, there is extrinsic evidence that the wheat crop was grown in the winter under irrigation. Compare: Morin, pp. 115, 252, 290.

> because of the contribution that we are obliged to make to his Majesty in the assessment of Royal Tributes...each of us is compelled to sow his own plot of wheat.[94]

The same necessity undoubtedly motivated the many other Indian pueblos growing winter wheat. The wheat production of Indian pueblos was characteristically small, but it had a certain importance in the social history of the communities.

The fiscal and legal records of the colonial period do not permit us to closely estimate the percentage of land under irrigation. But a contemporary source, cited by Morin, stated that one-fourth to one-third of the cultivated land in the diocese of Michoacán (including the Bajío as well as land to the south) was under irrigation around 1790.[95] Without proposing an alternative figure, we may still critique the accuracy of this estimate for the Bajío. We must first, however, remedy an omission: our local studies have not mentioned the subject of waterlifts or norias. Did norias significantly amplify the land under irrigation? It is clear that they did not.

The colonial Bajío was indeed dotted with many norias[96] that tapped the plentiful and accessible supply of groundwater. There is a record of the use of norias for irrigation near León in the eighteenth century.[97] On the Hacienda Otates, a large horse-powered noria supplied water to a tank of twelve cubic meters capacity from which it flowed to a network of irrigation ditches.[98] The Hacienda San Juanico near Celaya used a small noria to irrigate a plot of olive trees.[99] But one finds many indications that the principal use of norias was to provide water for horses and livestock during the long dry season; for example: the frequent association of norias with stock raising estancias,[100] their location on noncultivated portions of haciendas,[101] and the large noria troughs described in estate inventories.[102] The subdelegate of San Juan del Río, a district which was characterized by stock raising, explained that the many haciendas and ranchos lacking dams or ponds had built norias to supply them and their animals with water in the dry season ("para ellas

---

[94]AGN-T, Vol. 1124, Exp. 1, f. 23.

[95]Morin, p. 252.

[96]There were eighteen norias near Querétaro in 1848. Raso, Chap. 3. For references to norias in addition to those in the succeeding footnotes, see AGN-C, Vol. 73, Exp. 3, f. 20; AGN-T, Vol. 514, Exp. 1, Cuad. 2, f. 47 (illustration); Vol. 618, Exp. 1, Cuad. 3, f. 61; Vol. 629, Exp. 1, Cuad 4, f. 60; Vol. 666, Exp. 1, Cuad 1, f. 97; Vol. 1020, Exp. 1, f. 26v; Vol. 1353, Exp. 1, f. 69; ACM, leg. 834, Exp. 462, f. 14; AFC-CSC, leg "títulos" (Hacienda Valencia); AFC-CC, L-4, 1a pte, Cuad. 4, f. 22. For haciendas named La Noria, see AGN-T, Vol. 2367, Exp. 1; Vol. 2738, Exp. 33, f. 22; Vol. 2899, Exp. 45.

[97]Morfi, p. 381; AGN-T, Vol. 544, Exp. 3, f. 5v (land described as being "de noria").

[98]Morin, p. 244.

[99]AGN-C, Vol. 73, Exp. 3, f. 22v.

[100]AGN-C, Vol. 73, Exp. 3, f. 17 ("pila para los obejas"); AGN-H, Vol. 72, Exp. 9 (many references to norias, e.g., ff. 5v, 7v).

[101]AGN-T, Vol. 3401, Exp. 4, f. 14v ("en frente de la tienda" and "en el patio principal de la casa").

[102]ACM, leg. 821, Exp. 337, f. 11; leg. 830, Exp. 426, f. 16.

y sus animales").[103] Any use of the norias for irrigation was probably no more than incidental.

Our study serves to emphasize the close relation between irrigation and wheat production in the colonial Bajío. Any estimate of the extent of irrigation must begin with a consideration of the relative importance of wheat production. In Salvatierra during the mid-eighteenth century, wheat production fluctuated above and below the volume of maize production. In the Valle de Santiago the wheat harvest of 1750 almost exactly equaled the maize harvest, but in Celaya the volume of wheat harvested in that year was twenty-nine percent of the volume of maize, and in Salamanca (excluding the Valle de Santiago) it was less than 1 percent. During the period 1751-55, the average wheat production in León was only nine percent by volume that of maize.[104] The level of production in the alcaldía mayor of Querétaro was probably comparable to that of León, if we may judge from nineteenth century data. These proportions were, of course, subject to important regional and chronological variations, but a systematic analysis of tithe data is beyond the scope of this study. For our purposes, it is enough to note that, apart from two municipalities along the Lerma, wheat was a crop of distinctly secondary importance.

One can only conjecture as to the amount of the maize crop that was under irrigation. The tithe records do not distinguish between irrigated and unirrigated production, and documentary references to the irrigation of maize are few. But it should be borne in mind that haciendas storing water in systems of dikes plainly intended to use it for the irrigation of wheat; the use of storage basins for maize production or for growing lentils or garbanzos was purely incidental. The maize crop was irrigated extensively only in haciendas enjoying access to a perennial supply of water or an abundant reservoir, or engaging in the practice of medio riego. The scarcity of documentary evidence of such irrigation systems tempts one to conclude that they were of marginal importance, but a qualification is needed. In some cases, the paucity of documentary descriptions may reflect the limited importance of maize irrigation, but not its actual extent. The copious documents on the Hacienda Espejo near Querétaro refer most frequently to the irrigation of wheat and chile, but a single passage reveals that in 1799, more than twice as much land was irrigated for maize production as for wheat.[105] Similarly, we have seen that the eighteenth century records of the Valle de Santiago contain only a single reference to the irrigation of maize, but it is plausible to assume that the increase in maize production in the eighteenth century was assisted by irrigation of a portion of the maize crop in the spring or fall.

The irrigation of crops other than wheat and maize was clearly of small importance in the context of the regional economy. The question of the extent of land under irrigation thus turns on the amount of land devoted to wheat production. Roughly ninety percent of this land was irrigated. We may add to the wheat land a minor but still substantial share of the land devoted to maize cultivation. Morin's estimate of one-fourth to one-third of the cultivated land as being under irrigation is probably a reasonable approximation for the more favored parts of the Bajío, but within a broader regional context, it appears high. When H.G. Ward visited the Bajío in 1827, he was most impressed by the expanse of land left to dry farming and stock raising.

> I had pictured to myself a succession of haciendas, abundantly supplied with water for irrigation, and

---

[103]AGN-H, Vol. 72, Exp. 9, ff. 4v-5.

[104]ACM, leg. 847, 861, 836; MNA.AM-ACM, roll 776636 (1750); Brading (1978), pp. 69-70.

[105]AGN-T, Vol. 1353, Exp. 1, Cuad 2, ff. 38v-39.

> consequently smiling with verdure; and I was not a little disappointed at finding that the masses of cultivation, however considerable in their aggregate, were still lost in the immensity of the surrounding space.[106]

The irrigation systems that so impressed Humboldt did not, it seems, dominate the landscape.

The question of the extent of irrigation is of little interest in itself, but it is linked to a matter of capital importance: the level of agricultural productivity. The increase in the productivity of labor through the use of improved technology is an essential criterion of social progress. Morin argues that the extension of irrigation in the eighteenth century did bring a certain increase in the productivity of agricultural labor.[107] Our study of the valley of Querétaro supports this thesis. The irrigation systems near this major urban center were not simply devoted to wheat production but served generally to increase agricultural production through double cropping and the irrigation of maize, chile, and garden crops. Although the scanty documentary record of the region of Silao and Irapuato prevents a careful reconstruction of its agricultural history, we know that the agricultural economy of this region, closely linked with the mines of Guanajuato, experienced the most rapid expansion of any area in the eighteenth-century Bajío, and no fewer than twenty-three days dams lined the principal rivers by the end of the century.[108] Elsewhere one finds some scattered evidence to support the thesis of higher productivity; for example, the fact that the rental value of a day of water doubled in Celaya during the eighteenth century while the price of land probably remained relatively constant.[109] But we have shown that irrigation was always linked chiefly with the cultivation of wheat. It was thus employed to avoid fungal infestation and other climatic hazards of summer months rather than to raise the productivity of agricultural haciendas. On the whole, it appears that the application of irrigation to obtain higher productivity, so far advanced in the twentieth century, found only certain limited beginnings in the colonial period.

---

[106]Ward, II, p. 421.

[107]Morin, p. 252.

[108]Morin, p. 121. The legend of Figure 6.1 describes twenty-three "presas" and four "bordos" that may also have served the function of dams. The economy was based largely on the supply of maize to the Guanajuato mines. AGI Indiferente 107, tomo 1, 388. The scanty documentary record suggests that the waterworks were largely built in the course of the eighteenth century. AGN-M, Vol. 75, f. 225; Vol. 79, f. 16v; AGN-T, Vol. 1166, Exp. 1; Morin, p. 252. The absence of any mention of water in early land grants supports this inference. But see AGN-M, Vol. 22, f. 192v.

[109]See Brading (1978), pp. 83-85.

6
Technology

The technology of colonial Mexico, though an essential part of its culture, has long eluded historical research. It is only recently that a small literature, including the distinguished work of Roberto Moreno, has begun to appear. The legal documents that form the material of most historical investigations ordinarily yield a very thin harvest of data on technology. But the water cases of the Bajío have proved to be a more productive source of information than most documents of this sort. Here, we will survey generally the methods of irrigation, study four irrigation systems in some detail, and then consider the topics of water measures, leveling techniques, and professional involvement in water works.

METHODS OF IRRIGATION

Masonry storage dams were the principal instruments of irrigation on minor arroyos and the small rivers of the northwest Bajío. The most impressive series of storage dams was distributed along the length of the Rivers Silao and Guanajuato. The floodgates of the dams were closed until their reservoirs were filled; water was then released to the next hacienda downstream. The largest of these dams on the Hacienda Arandas was said to create "una gran laguna," although its exact dimensions cannot be established. A map prepared in 1792 provides a crude but comprehensive overview of the irrigation systems on these rivers (Figure 6.1). The panorama of the region of León, described by Brading, presents some similarities to that of Irapuato, although there were fewer irrigated haciendas on the River Turbio.[1] The Hacienda de la Lagunilla possessed

> A masonry dam three hundred varas wide, with nine pillars in the middle and others on the sides, twelve varas high in the center, with foundations said to be buried in places about seven varas, with a thickness of at least two varas.[2]

---

[1]Brading (1978), p. 120.

[2]AGN-C, Vol. 272, Exp. 5. (The dam was valued at 3,000 pesos; the total value of the hacienda was 15,592 pesos.)

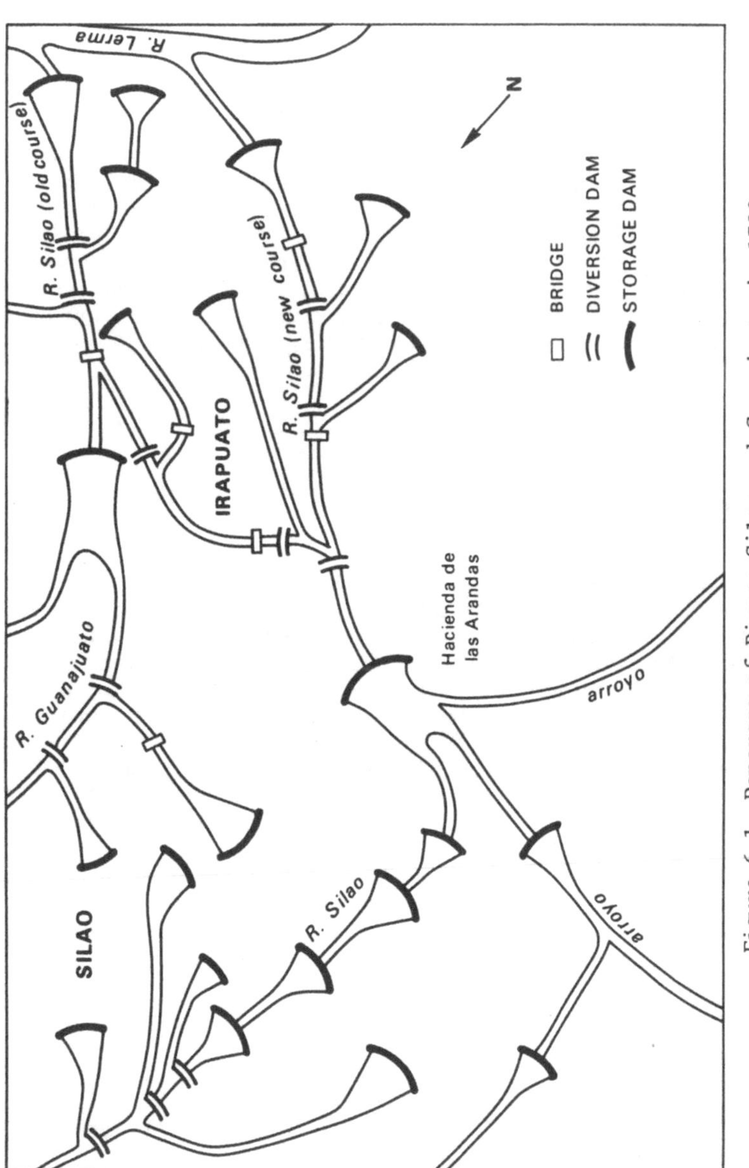

Figure 6.1 Panorama of Rivers Silao and Guanajuato in 1792

At the other end of the Bajío, many haciendas of San Juan de Río also relied on storage dams. The largest was on the Hacienda Llave, which was described in 1794 as having "a famous dam four hundred varas wide."[3] On minor arroyos, storage dams were built either of masonry or of earth with masonry floodgates.[4] The Hacienda Ojo Zarco near Apaseo had two masonry dams; one 135 varas long and twelve and one-half varas high, and the other less than one-half this size.[5] The earthen dam of the Hacienda San Lucas, located in the hills south of Celaya, was calculated to have 32,570 cubic varas of material. The wooden floodgates, placed between masonry frames, consisted of "fifteen planks of *ahuehuete* two varas long, two-thirds vara wide and one-third vara thick..."[6]

Diversion dams, though found in the northwest Bajío, were of particular importance in the intensely irrigated land west of Querétaro and along the Rivers Laja and Lerma.[7] A common form of construction consisted of masonry pillars anchored in the riverbed and joined by gates of removable wooden planks (*tablones*). The earliest description of a dam that I have found was of this type. The dam, built in 1620 to channel water to the villa of León, possessed

> Two masonry pillars in the arroyo with their abutments that project out narrowing the course of the arroyo as much as possible so that an arch can be placed above them.[8]

We have a meticulous drawing of a similar dam built in 1760 near Irapuato to divert water alternatively to one of two channels of the River Silao (Figure 6.2).

Other diversion dams consisted of earthen barriers placed within the riverbed that served to raise the water to the level of masonry headgates located at one side of the river. The headgates, rather than the dam itself, regulated the flow of water to the fields and storage basins. Despite the damage caused by summer rains, the dams could be annually repaired and put back into use. The Haciendas Cieneguilla and Petaca in San Miguel El Grande possessed extensive irrigation systems based on this sort of dam (Figures 6.3 and 6.4).[9] The series of earthen dams north of Celaya were probably of the same basic type.[10]

With favorable terrain, there was actually little need for a dam. Although López de Peralta claimed to have spent 20,000 pesos in building the irrigation system of his hacienda in Salvatierra, the dam at the head of the canal was described in the eighteenth century as consisting of no more than "una cerquilla a parecer de tierras y piedras

---

[3]AGN-H, Vol. 72, Exp. 9, f. 8.

[4]AGN-T, Vol. 1020, Exp. 1, Cuad. 2, f. 37v; Vol. 1041, Exp. 2, f. 46; Vol. 3401, Exp. 1, Cuad. 1, f. 19.

[5]AGN-C, Vol. 1298, Exp. s/n, ff. 25-35 (Ojo Zarco).

[6]AGN-T, Vol. 3401, Exp. 1, Cuad. 2, f. 4.

[7]In addition to other descriptions of diversion dams cited herein, see AGN-T, Vol. 383, Exp. 4, ff. 3-12; AGN-M, Vol. 39, f. 203v; Appendix 1.

[8]AHML, caja 1620, Exp. 27.

[9]AGN-T, Vol. 1168, Exp. 3, Cuad. 2, ff. 18-34; Vol. 1169, Exp. 1, Cuad. 5, f. 73v; Vol. 1175, Exp. 4.

[10]AGN-T, Vol. 2071, Exp. 1, ff. 1, 47.

sueltas."[11] In some circumstances, a canal could be incised in the riverbank to the level of the water and protected with large masonry headgates. I have found only one description of such a system, but the frequent references to *tomas* or sacas in Salvatierra and Celaya without any accompanying mention of dams leads one to infer that there were other systems of this sort.[12]

Removable floodgates were needed for distinct reasons and their use followed various patterns.[13] Haciendas sharing the flow of a small stream were often bound by agreements with downstream users to periodically release water.[14] On a perennial stream, gates might be raised to avoid flood hazards during summer months and be lowered only during periods of irrigation in the winter. Landowners neighboring the Hacienda Santa Rita forced the owner to agree not to lower his floodgates on the Lerma before October 15 of each year.[15] The diversion dams on seasonal streams served to channel water during the summer months to storage basins; the gates were presumably lowered for this purpose and kept in place only until the basins were full. The use of headgates on canals followed more obvious patterns. Under favorable circumstances, they were opened and closed as needed for irrigation or the storage of water, but, just as floodgates were opened during high water, a headgate placed in a riverbank might be closed to avoid flooding.[16]

The haciendas of colonial Mexico employed a variety of water storage facilities. In stock-raising country one finds references to simple ponds (jagüeyes) located in a depression in the ground and often diked on one side. Occasionally, the ponds were lined or partially enclosed with masonry to increase their storage capacity.[17] Adjacent to the hacienda buildings, there were often masonry boxes or tanks (*pilas, albercas*) to store water for domestic use.[18] But the wheat haciendas of the Bajío employed characteristic water storage facilities, described as *bordos* or *cajas de agua*, which I have translated as "storage basins." They consisted of earthen dikes, usually of modest height but occasionally as much as five or six meters high, enclosing an expanse of relatively level ground, frequently ten hectares or more in size (Figure 6.5).[19] Masonry gates regulated the flow of water to and from the storage basins. The gates that I have seen have two parallel grooves on each side into which small logs or planks can be fitted to create two

---

[11]AGN-T, Vol. 988, Exp. 1, f. 86v.

[12]E.g., AGN-M, Vol. 16, f. 117v; Vol. 33, f. 258v; AGN-T, Vol. 1247, Exp. 1, Cuad. 1, ff. 6, 35; Vol. 2071, Exp. 1, f. 1; Vol. 2809, Exp. 14.

[13]AGN-T, Vol. 1170, Exp. 1, ff. 242, 322; Vol. 3618, Exp. 1, Cuad. 3, f. 38. A survey of dams in the Valley of Mexico similarly contains numerous references to wooden floodgates. Cuevas Aguirre y Espinosa, passim.

[14]AGN-M, Vol. 79, f. 17 (Irapuato); AGN-T, Vol. 192, Exp. 1, f. 10.

[15]See Appendix 1.

[16]See the discussion below of Haciendas Molino, Mendoza, and Espejo.

[17]See AGN-H, Vol. 72, Exp. 9, ff. 1-21.

[18]E.g., AGN-C, Vol. 1298, Exp. s/n, f. 27 (Hacienda Ojo Zarco).

[19]AGN-T, Vol. 1020, Exp. 1, Cuad. 2, f. 33v; Vol. 3618, Exp. 1, Cuad. 2, f. 7; AGN, Fomento Caminos, Vol. 90, f. 14v; SRH, Celaya, folder labeled "generalidades 'Rio Laja.'"

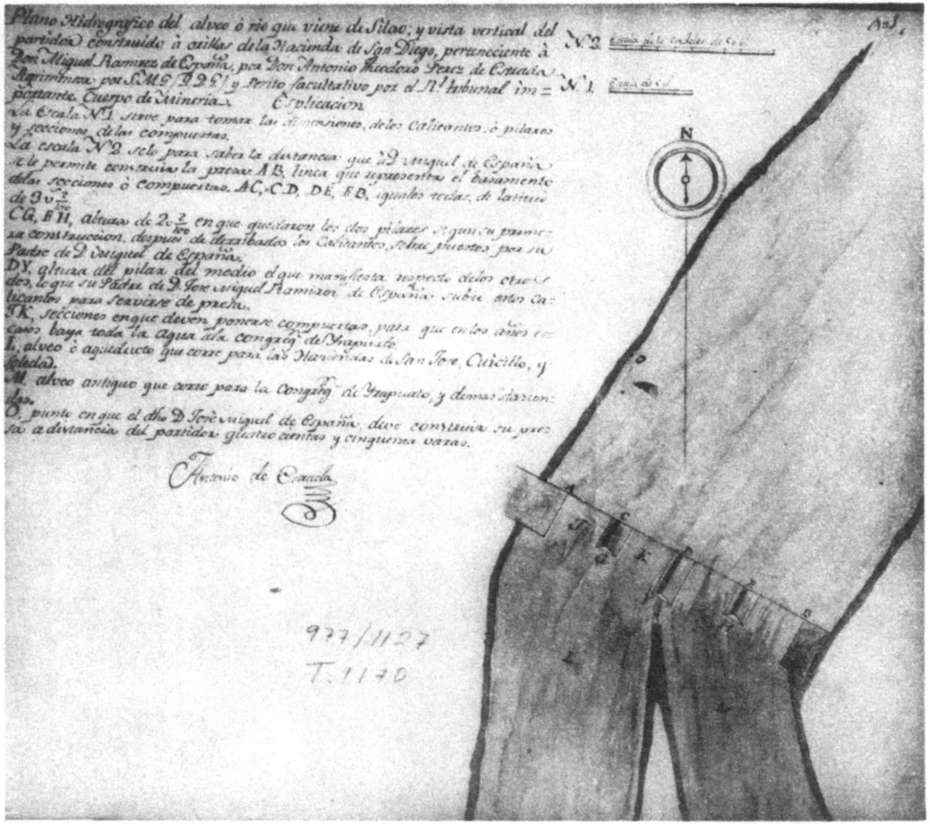

Figure 6.2 Diversion dam, Irapuato, around 1760
(Source: AGN-T Vol.1170, exp.1)

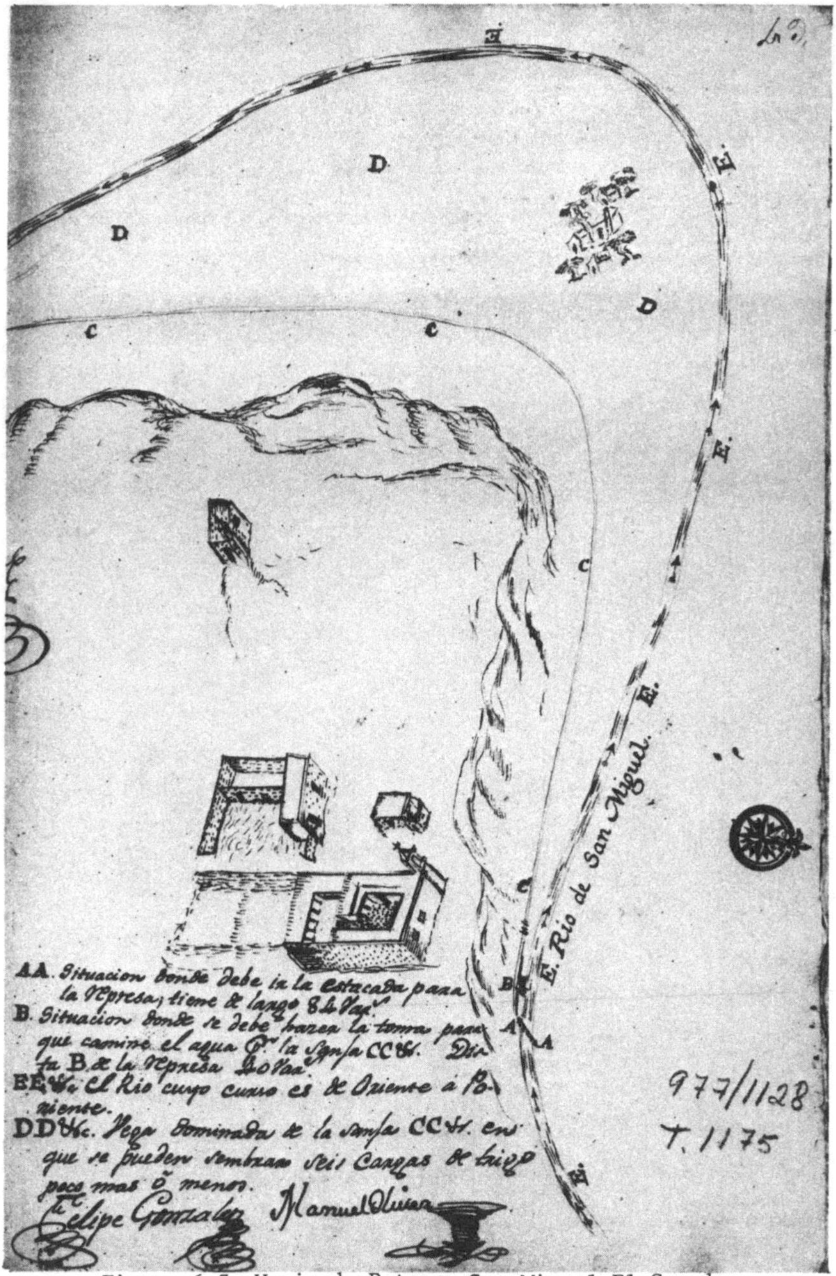

Figure 6.3 Hacienda Petaca, San Miguel El Grande
(Source: AGN-T, Vol.1175, exp.4)

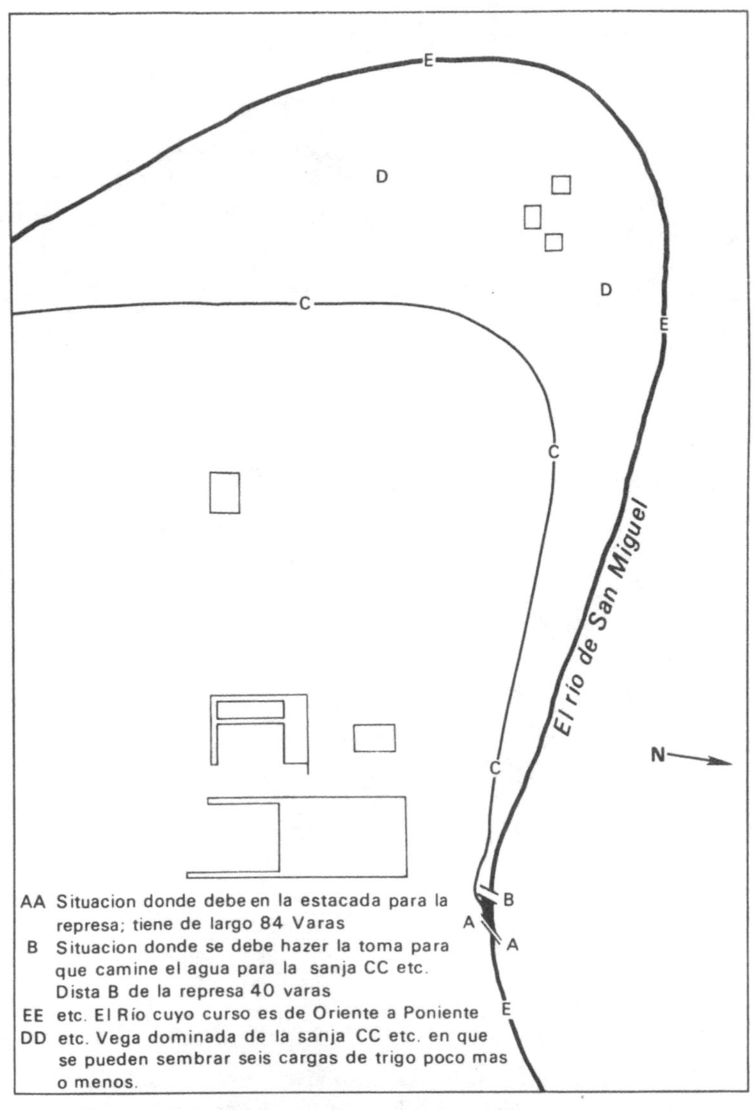

Figure 6.3 Hacienda Petaca, San Miguel El Grande

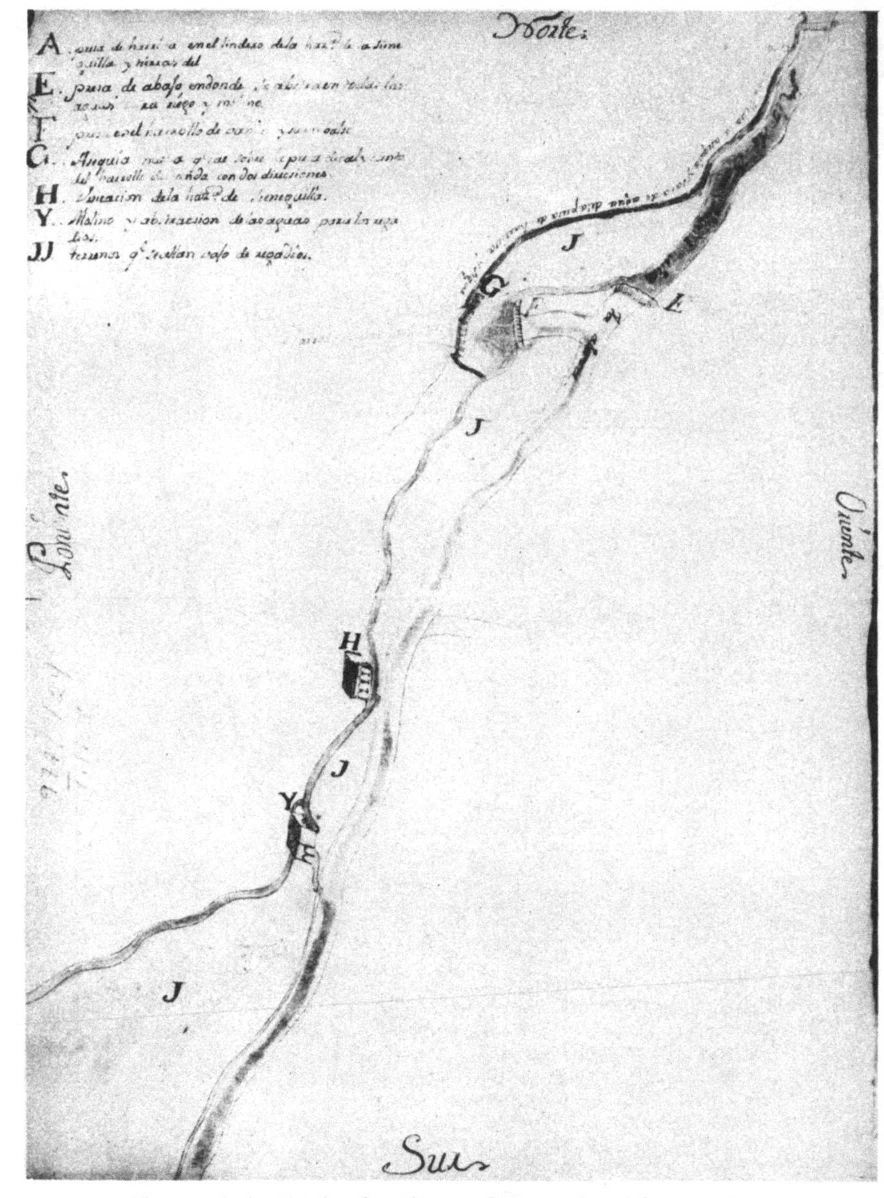

Figure 6.4 Hacienda Cieneguilla, San Miguel El Grande (Source: AGN-T, Vol 945)

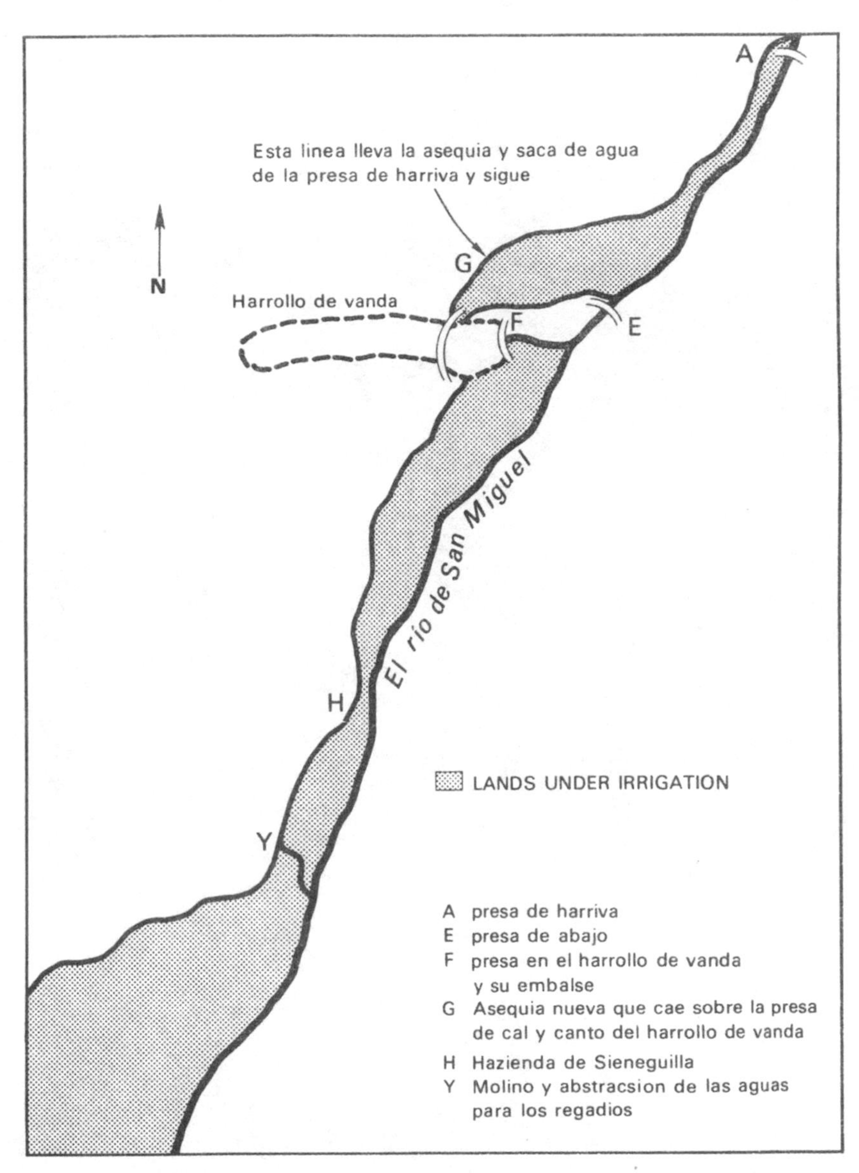

Figure 6.4 Hacienda Cieneguilla, San Miguel El Grande

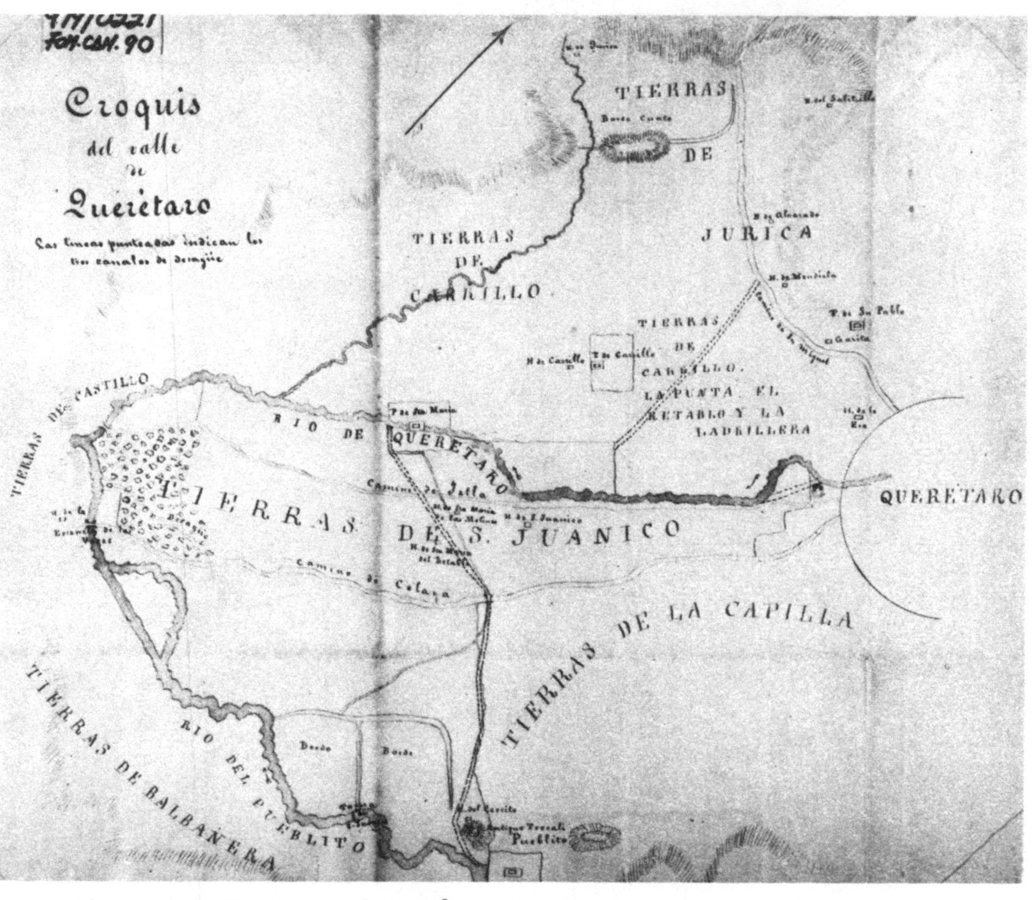

Figure 6.5 Environs of Querétaro in 1865
(Source: AGN-Fomento Caminos,3650)

roughly formed walls. The space between the walls can then be filled with earth to make the gate sufficiently watertight. Although storage basins of this description usually received water diverted from dams, two haciendas near Querétaro built large storage basins at the foot of small arroyos where they served to contain the intermittent flow.[20]

Water was generally conducted to fields or storage basins through simple ditches. The fine texture of the Bajío soil, which commonly has forty or fifty percent clay content, prevented excessive loss of water in conduction, and the level terrain presented few obstacles.[21] But one does find scattered references to specially constructed conduits.[22] The Hacienda San Isidro in Acámbaro received water through wood *canoas*.[23] The canals of Gogorrón and the mayorazgo in Salvatierra traversed some uneven ground that required the construction of small aqueducts and masonry conduits (*targeas*).[24] The need for thrifty use of the modest perennial supply of water in Querétaro led to the construction of masonry conduits not only within the town but also on canals leading to the wheat farms at the periphery.[25]

Masonry fixtures were placed at intervals within the ditches to regulate the distribution of water. In all areas of the Bajío one finds innumerable references to masonry irrigation gates (marcos).[26] In the late eighteenth century, the farms in the Valle de Santiago received water through "a gate one-third of a vara wide placed level with the bottom of the ditch and solidly constructed."[27] A system of ditch dams and irrigation gates is described in the environs of Salvatierra.

> Those who irrigate with the water of the canal have various open gates with which they irrigate the adjacent plots and others...and next to these gates are placed certain obstacles or small dams to raise the water to flow through the gates since the water does not have adequate height in its normal course.[28]

The larger systems of allocation involved the use of irrigation boxes with locked gates.[29]

---

[20]MNA.AH Gómez de Orozco, leg. 1, ff. 235-43 (Hacienda LaGriega) and map 27 (Hacienda Jurica). See also AGN-T, Vol. 1020, Exp. 1.

[21]See footnote 50, Chap. 5.

[22]AGN-T, Vol. 185, Exp. 1, Cuad. 1, f. 15, Cuad. 2, f. 14; Vol. 3401, Exp. 1, f. 19; AGN-C, Vol. 1298, Exp. s/n, f. 35 (Ojo Zarco).

[23]AGN-T, Vol. 534, 2a pte, Exp. 1, Cuad. 1, f. 1.

[24]AGN-R, Vol. 8, Exp. 1; AGN-T, Vol. 1247, Exp. 1, Cuad. 1, ff. 76, 360.

[25]MNA.AH, Fondo Franciscano, Vol. 92, f. 73.

[26]E.g., AGN-T, Vol. 2959, Exp. 141, f. 4v; Vol. 2951, Exp. 31; AGN-M, Vol. 70, f. 109.

[27]AGN-A, Vol. 97, Exp. 2, f. 10.

[28]AGN-T, Vol. 988, Exp. 1, Cuad. 3, f. 144. Also AGN-T, Vol. 764, Exp. 1, Cuad. 3, f. 10; MNA.AM-Qro, roll 23, first doc., f. 305.

[29]AGN-A, Vol. 108, Exp. 1, and Exp. 2, f. 54; Vol. 169, Exp. 20.

Curiously little attention was paid to the problem of drainage. Although the Valle de Santiago had a drainage canal, I have found only one other reference to a drainage ditch.[30] The neglect of this aspect of irrigation, so important in modern systems, appears to reflect the relative abundance of land and scarcity of water.

Simple, labor-intensive methods of irrigation were sometimes employed on small water courses. The subdelegate of San Juan del Río reported in 1794 that he had sought to encourage the construction of small dams of earth and stone because they could be easily constructed ("por que es facil su applicacion"). The dams were rebuilt annually or as often as necessary ("se forman anualmente, y quantas vezas son necesarias").[31] The Indian pueblo of San Francisco del Rincón had a dam of pressed earth, reinforced with stakes, that had to be periodically rebuilt.[32] The productive wheat haciendas along the Brazo de Moreno in the Valle de Santiago also used techniques that might appear to be somewhat primitive. The dams were typically anchored with large timbers called "mules"; the bulk of the dam was woven out of stakes, poles, willow branches, and turf and reinforced with pressed earth.[33] These dams, called *bolsas*, could be rebuilt annually at small cost; a landowner complained that the reconstruction of one dam cost seventy pesos, apparently an unusually large expense. The fact that the dams leaked was a kind of advantage because it minimized conflicts with downstream users. A judicial decree actually ordered the Hacienda Sotelo to build dams only of the bolsa type so that enough water would pass to a flour mill downstream.[34]

SAMPLE IRRIGATION SYSTEMS

Surviving documents contain good descriptions of four interesting irrigation systems of the colonial Bajío. Two of these, those of the Hacienda Mendoza and the Hacienda Espejo, are still intact so that I was able to supplement the documentary descriptions by personal field observation.

Hacienda Mendoza

At the base of a modest arroyo north of Salamanca in the former Hacienda San José de Mendoza stands an imposing wall of masonry extending about 193 varas from either side of the canyon wall (Figure 6.6). It is about six varas high toward the center and five and one-half varas thick at the base, rising to a narrow ridge just wide enough to walk on. The dam was built in 1740 by a rich miner from Guanajuato, Francisco

---

[30]AGN-T, Vol. 432, Exp. 1, f. 12; AGN, Vínculos, Vol. 166, f. 156v (drawing).

[31]AGN-H, Vol. 72, Exp. 9, ff. 3, 13.

[32]AGN-T, Vol. 1124, Exp. 1, f. 21. Also AGN-T, Vol. 822, Exp. 2, f 15v.

[33]AGN-T, Vol. 2963, Exp. 49, f. 1; Vol. 1113, Exp. 1, Cuad. 1, f. 19v, Cuad. 5, f. 6v, 24; Vol. 1107, Exp. 1, Cuad. 2, f. 23.

[34]AGN-T, Vol. 1113, Exp. 1, e.g., Cuad. 3, f. 14v, Cuad. 6, f. 41; Vol. 1114, Exp. 1, f. 73, 221.

Figure 6.6 Dam of Hacienda Mendoza in 1981

Yguerategue.[35] Its excellent condition today is a tribute to the craftsmen who constructed it. In the middle of the dam is a narrow gate about two feet wide with two parallel grooves about three inches wide on each side. Small logs have been placed in the grooves and the intervening space filled with dirt. The small spillway at the base of the gate is no longer used. Behind the dam is a surprisingly small lake; the steep gradient of the arroyo limits the size of the storage basin.

The dam was built in about seven months; the foundations were excavated by December 28, 1739, and the last load of lime delivered on June 19, 1740. The notes of the supervisor of construction, preserved in the records of a lawsuit, reveal how he proceeded to employ short-term labor and coordinate the delivery of stone, sand, lime, and wood (since the lime was produced in an oven near the site, large quantities of firewood were required). Some of these expenses can be estimated with the help of cost figures from records of the nearly contemporaneous construction of a dike in Irapuato.[36] About 1,800 *carretas* (a measure of volume much smaller than the typical haul) of stone and sand were delivered. At three reales a carreta, the expense would be about 675 pesos. In thirteen burnings, the oven produced 2,313 fanegas of lime, which had a market value of about 580 pesos. The employment of teamsters to haul the stone, limestone, and firewood was an important expense; the notes, though not fully legible at this point, appear to indicate an expenditure of about 400 pesos.

A throng of peons, hired mostly for brief and irregular periods, put in somewhat less than 3,000 man days of labor, the equivalent of about twenty laborers working full time for seven months. Fifty-five *peones alquilados* were hired for periods ranging from a few days to sixty-two days. Twenty-eight salaried peons worked for longer periods of as much as ninety-five days. And thirty-seven *muchachos* were hired at one-half the normal rate of pay for periods of up to ninety-five days. At a rate of one real a day for the muchachos and two reales for the other peones, the labor cost amounted to only about 550 pesos.

Fifteen masons worked on the dam at various times and contributed over 400 days of labor for a total cost of 495 pesos. A mason from Celaya was employed at a rate of one peso two reales per day; others were evidently paid somewhat less. The supervisor's son was paid 644 pesos for his services; the supervisor himself sued for payment at a rate of five pesos per day. If we may assume that he recovered around 1,000 pesos and add in other expenses appearing in the notes, the total cost of the dam would be on the order of 5-6,000 pesos.

The supervisor of construction of the dam was related to the owner by ties of *compadrazgo* but evidently possessed no special training or experience for the job. The site was poorly chosen, and the dam failed to achieve its purpose of irrigating a wheat crop on the rich bottom land of the hacienda. An appraisal of the hacienda in 1742 stated that the dam was very useful for watering livestock but served to irrigate only a very small part of the hacienda's sixteen caballerías of agricultural land. The dam had inadequate storage capacity, and much water was absorbed in the sandy soil at the base of the arroyo as it was conducted to the agricultural land a couple of kilometers away. The dam, nevertheless, represented a substantial part of the value of the hacienda. The 1742 appraisal gave it a value of 9,500 pesos and the entire hacienda a value of 39,000 pesos. A second appraisal, however, put the value of the dam at 6,000 pesos.[37]

---

[35]AGN-T, Vol. 274, Exp. 1, Cuad. 6.

[36]AGN-T, Vol. 1166, Exp. 1, f. 15 et seq.

[37]AGN-T, Vol. 275, Exp. 1, Cuad 2, f. 52, Cuad. 3, f. 117.

Hacienda del Molino

We have already encountered the irrigation systems of the Hacienda del Molino in our discussion of the swamp water and flood control problems of Celaya. The hacienda lay at the southern edge of the city. In 1789, the owner sought to divert water from the Laja as well as to tap drainage from various sources north of the city.[38] Rather than building a dam, he incised into the side of the riverbank a canal that extended from a point north of the city seven and one-half kilometers southwest to his hacienda. The project proved to be successful in irrigating the hacienda but, as we have seen, offensive to the residents of Celaya. The builder of the canal and his successor were plunged into litigation that produced a very skillful drawing of the system in 1806 (Figure 6.7).

The head of the canal was excavated to a depth of one vara below the level of the riverbed, and massive masonry headgates, reinforced with earth and stone, were placed on the bank to regulate the flow of water.[39] The headgates could be opened or closed as needed by the use of planks (tablones) placed between two masonry frames; these gates were said to admit over four *bueys* (four square varas) of water, but were apparently later reduced in size to one square vara.[40] The initial portion of the canal was necessarily narrow and deep; the Laja cuts a trench five to ten meters deep to the northeast of the city. The first 200 varas of the canal were in fact described by a term (*ademe*) usually applied to mining shafts. Farther on, the canal opened up and assumed a width of as much as ten varas.[41] The best descriptions of the headgates and canal are found in the following passages.

> Headgates constructed with masonry at great expense, placed firmly and solidly at the edge of the river bank, with a pair of gates for its use and a ditch or aqueduct extending behind it on the western side...
>
> ***
>
> Consists of two gates secured by three pillars connected by arches and domes which form bridges and from the top pillars extend two long and thick walls and in the canal wooden supports extend twenty varas back of the headgate.[42]

The headgates were closed during high water and opened when the water subsided to manageable levels.[43] The depth of the headgates assured that the system would capture

[38]AGN-T, Vol. 2767, Exp. 3, Cuad. 3, f. 1v.

[39]AGN-T, Vol. 2071, Exp. 1, Cuad. 3; Vol. 2072, Exp. 1, f. 16; Vol. 1390, Exp. 3, f. 29v.

[40]AGN-T, Vol. 2072, Exp. 1, Cuad. 5, f. 13 (the course of the lawsuit indicates that the expert's recommendations were adopted).

[41]AGN-T, Vol. 2767, Exp. 3, Cuad. 2, f. 1v.

[42]AGN-T, Vol. 2767, Exp. 3, Cuad. 2, f. 27, Cuad. 5, f. 16.

[43]AGN-T, Vol. 2767, Exp. 3, Cuad. 5, f. 3v.

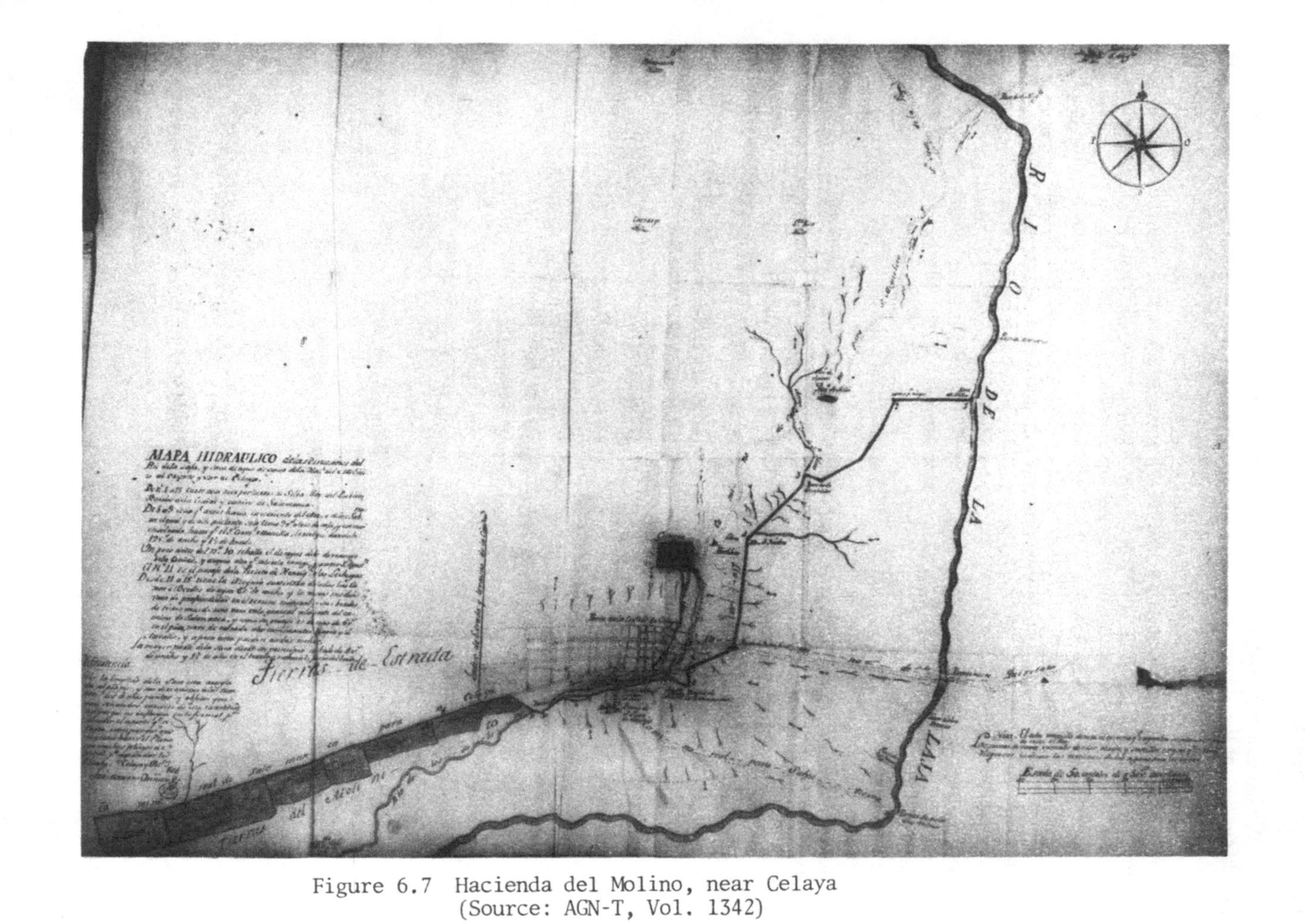

Figure 6.7 Hacienda del Molino, near Celaya
(Source: AGN-T, Vol. 1342)

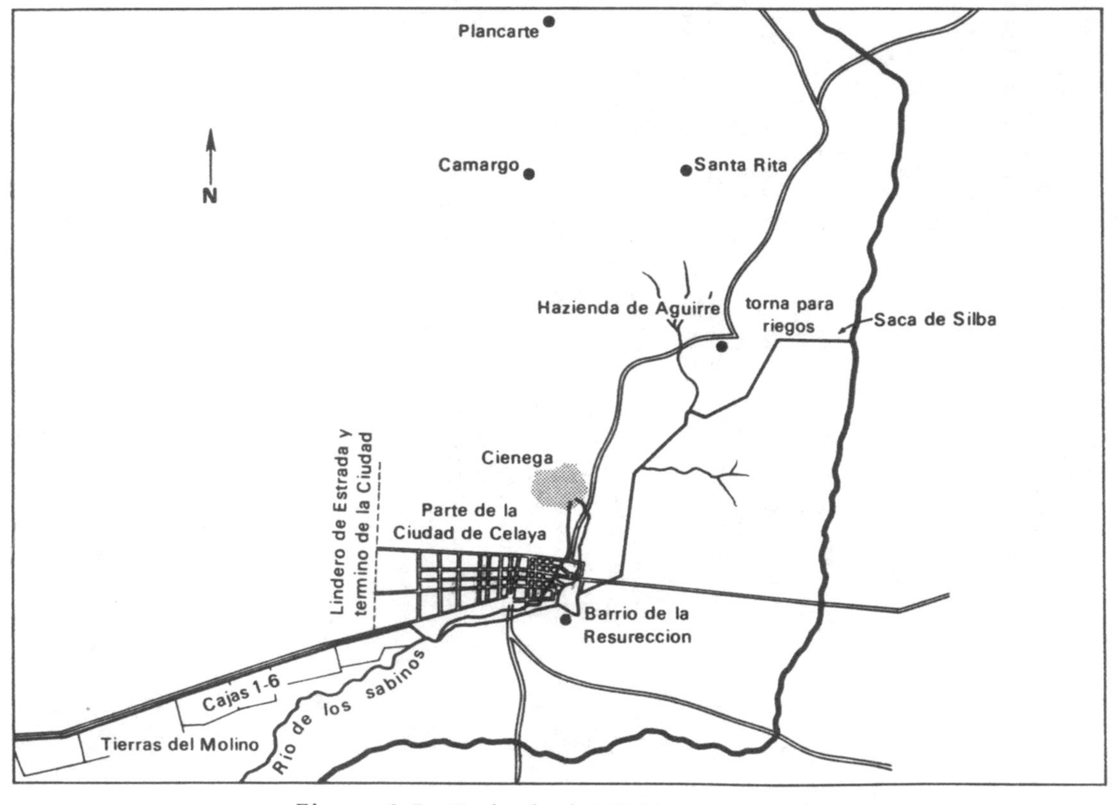

Figure 6.7 Hacienda del Molino, near Celaya

some of the small perennial flow of the river, and the initial stretch of the canal actually tapped groundwater supplies.[44]

The canal proceeded directly west and then cut to the southwest. After a distance of about two and one-half kilometers, it joined a kind of gully formed by floodwaters and surface runoff. As it proceeded toward the city, the canal was described as being generally not less than four varas wide and one and one-half varas deep. South of the city it was augmented by water draining from the swamp and irrigation systems.[45] At the point shown by an "11" on Figure 6.7 stood a small masonry dam, built many years before by a previous owner, that was used to divert water to the hacienda.[46] The dam consisted of three pillars, joined by removable wooden planks, that created a pond about 150 varas long. The gates could be adjusted to provide the appropriate flow of water ("compuertas corredizas para usar de ellas a proporcion de la necesidad o circumstancias de la creciente").[47] The water was conducted through the hacienda along a shallow ditch to six basins that may have been used either for cultivation or water storage--most likely for both purposes.[48]

Hacienda Castillo

The Hacienda Castillo had long irrigated from headgates located near the confluence of the River Querétaro and the River Pueblito. In 1787, the owner of the hacienda, Tomás Ecala, set about to increase his irrigated acreage by building two earthen diversion dams and larger headgates. These improvements unfortunately flooded neighboring properties and reduced the flow to downstream haciendas, but the ensuing litigation produced a drawing of the irrigation system in 1799 that presents an interesting study in colonial technology (Figure 6.8).[49]

The earthen diversion dam of the hacienda indicated by the number "1" is located just beyond the juncture of the two rivers; the dam indicated by the number "2" is located 230 varas upstream on the River Pueblito. Ecala is said to have first built the dams of material that was quickly carried away and to have rebuilt them of compacted earth, reinforced with stakes and interwoven with branches. At least one of the dams was placed against a row of loose stones. The dam number "1" was twenty-two varas long and thirteen varas wide and of unspecified height; dam number "2" was fourteen varas long, ten varas wide and two and one-third varas high.[50] A third earthen diversion dam, identified by the letter "O" is shown downstream. Masonry dams were, however, equally common on the river. The masonry dam designated "P" was of the common pillar and

---

[44]AGN-T, Vol. 1390, Exp. 3, f. 32v.

[45]See legend, Figure 6.7.

[46]AGN-T, Vol. 2071, Exp. 1, Cuad. 1, f. 9v.

[47]AGN-T, Vol. 1390, Exp. 3, f. 2, 4v and 24v.

[48]AGN-T, Vol. 2072, Exp. 1, Cuad. 5, f. 21.

[49]AGN-T, Vol. 3618, Exp. 1, Cuad. 3, f. 8v. A description of the drawing in Volume 3618 contains most of the information in this section.

[50]Ibid, Cuad. 1, f. 11v-18.

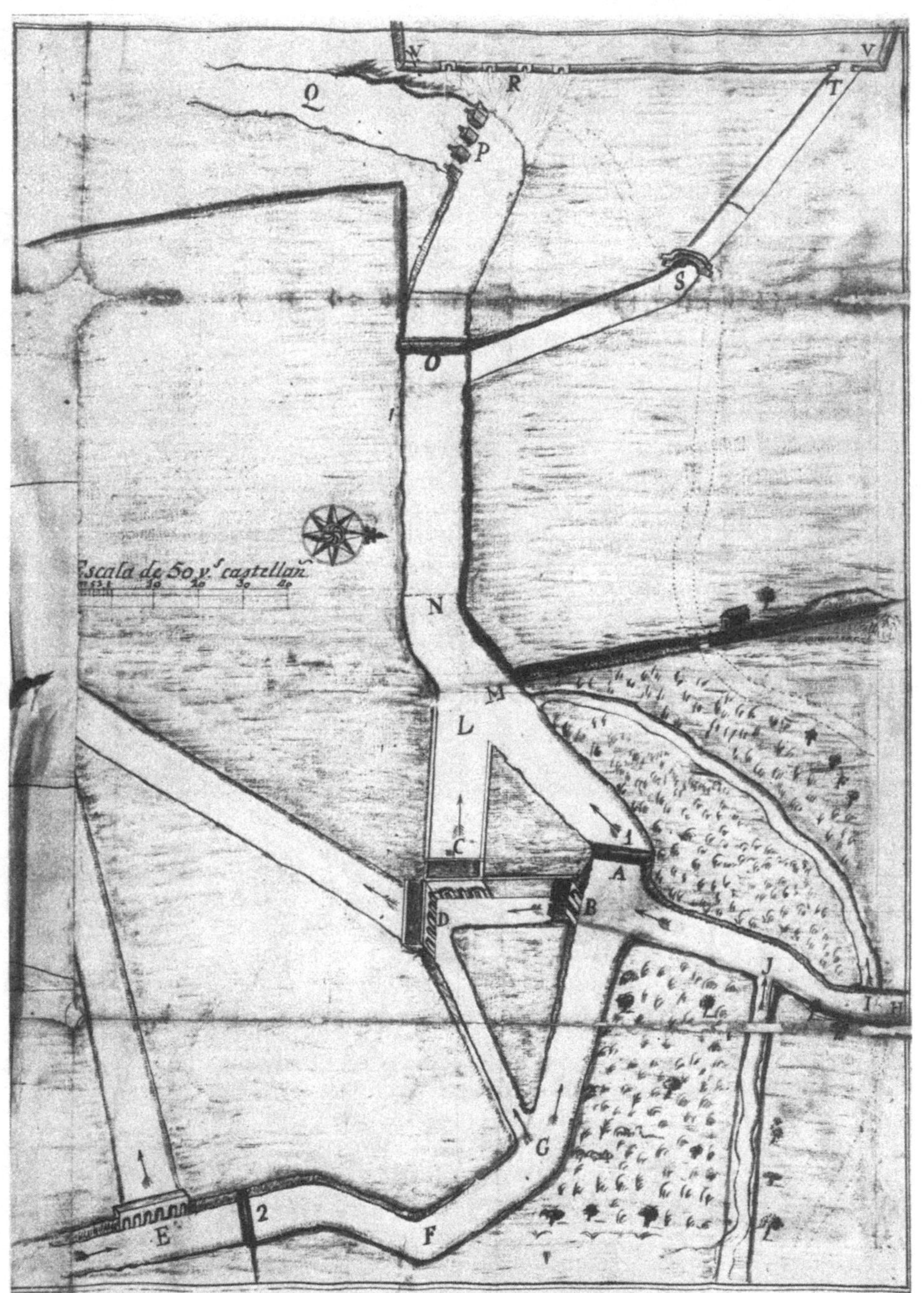

Figure 6.8 Hacienda Castillo with irrigation system (Source: AGN-T, Vol. 2499)

gate form of construction, and other masonry dams lay beyond the area of the drawing both upstream and downstream.[51]

The Hacienda Castillo had formerly received water only through the headgate "B." Ecala dug a new channel at point "G" and built the additional headgates "D" and "C." The headgate "D" admitted water to the canal leading to the hacienda's storage basins; the headgate "C" released water downstream when it was not needed. Upstream, Ecala built a fourth headgate that admitted water to a secondary canal leading directly to the principal canal. The arched openings on these headgates were generally one or one and one-half varas wide, but the original arched headgate "B" was two and one-half varas wide. A pillar placed next to this headgate "B" expanded its capacity by creating a gate one vara wide.

The water channeled to the hacienda by this system of diversion dams and headgates was stored in four storage basins described as

> four large dikes, holding ponds or deposits, thousands of varas in circumference and six or seven in height in which the water is deposited to provide up to three waterings for the wheat.[52]

At the top of the drawing is shown part of the irrigation system of the Hacienda La Punta. Two dams, previously described, diverted water to a storage basin located close to the river. The length of the eastern side of the storage basin between points "V" and "V" was 400 varas. If we may assume that the basin had a roughly square shape, it had an area of about sixteen hectares. The hacienda had a second storage basin beyond the limits of the drawing. While the Hacienda Castillo regulated the flow of water to its storage basins by means of headgates on the river, the Hacienda La Punta relied on the irrigation gates of the storage basin itself at points "T" and "P" and on the wooden gates of the dam "P" which could be put in place or removed as needed.

Hacienda Espejo

The irrigation system of the Hacienda Espejo, built in the mid-eighteenth century,[53] represents a triumph of colonial engineering skill. The dam stands where the arroyo Mandujano cuts through a low ridge of hills and plunges into a small rock gorge above the valley floor west of Querétaro (Figure 6.9). The dam is not excessively large; it consists of a thick masonry wall about 125 yards long, occupying the main portion of the arroyo, with extensions angling off in the direction of higher ground at either side. The dam rises about five meters above the present level of the lakebed and almost fifteen meters above the canyon floor on the far side. The top two meters of the dam are a later construction, added in modern times to offset the effect of siltation. Despite its modest size, the dam is located in such a strategic place that it creates a lake about four

[51] Ibid, Cuad. 3, f. 36-38.

[52] Ibid, Cuad. 2, f. 7.

[53] AGN-T, Vol. 1354, Exp. 1, Cuad. 1, f. 66-69. The dam was built by Pedro Bernardino de Primo y Jordán who died in 1764. Also AGN-C, Vol. 120, Exp. 2.

Figure 6.9 Dam of Hacienda Espejo in 1981

kilometers long and about one kilometer wide,[54] surrounded by gentle hillsides. The water is drawn off for irrigation through a tunnel about 900 meters long that extends under a hill on the eastern side of the lake about one kilometer from the dam itself. The tunnel now in use lies parallel to the original tunnel, which once received water from a point level with the bottom of the dam. The older tunnel is intercepteed about every 100 meters by vertical shafts that permitted entrance for maintenance and repairs.

At the base of the dam are arched openings that probably once contained floodgates. A smaller archway half way up the dam opens into a kind of spillway. In the eighteenth century, the dam's reservoir filled up in all except very dry years and overflowed into the canyon in rainy years. The floodgates may have been opened on these occasions; the release of water at the base of the dam would have curtailed the process of siltation. But the archways are now closed with masonry, and excess water is allowed to spill over the top and around the side of the dam.[55]

The discharge of water from the reservoir was controlled by a headgate at the entrance to the tunnel and by an irrigation box and water dividers at the other end (Figure 6.10). The headgates were rebuilt in 1810 to incorporate improvements recommended by a professional surveyor (*agrimensor*). Today one can see a shaft twenty-three inches across having vertical slots two inches wide. The slots have been very finely cut into a hard stone, especially selected for the purpose, that is placed in the masonry at either side of the shaft. The smooth and precise line of the slots allowed well-fitted planks (*tabletas* or *barrotes*) to be routinely raised or lowered. The shaft was enclosed and access was restricted by a locked door. Regularly spaced holes that once probably supported a ladder are located on either side of the shaft leading to the bottom ten meters or so below. A description of the recommended design of the gate in 1810 indicated that the slots would be fitted with a succession of planks, one on top of the other, that could be raised individually and somehow locked into position.

> The aforementioned room locked by two keys is very important, but it can be reduced to a staircase with only a door and some openings that give access to the aforementioned planks (barrotes) so that without leaving the staircase one can remove or release them and fasten them by means of a ring placed in the structure that functions with another ring embedded in lead in the masonry to which a lock can be attached. In this manner, the upper plank may be secured, and when the water falls to this level, it may be removed and then the next and so on to the last.[56]

At the end of the tunnel, the water entered an irrigation box that controlled the flow into a flour mill (milling was said to be the most profitable business of the hacienda around 1810) and to several masonry water dividers leading to the principal ditches. Both the irrigation box and the dividers were fitted with locked gates.[57]

---

[54]My estimate is based on pacing off the length and breadth of the lake when it was dry. The CETENAL map, however, shows the lake to be less than three kilometers long and one kilometer wide.

[55]Ibid, Cuad. 1, ff. 107-11v.

[56]Ibid, Cuad, 2, f. 32. Also AGN-T, Vol. 1354, Cuad. 2

[57]Ibid, Cuad. 1, ff. 109-21, Cuad. 2, ff. 32-41.

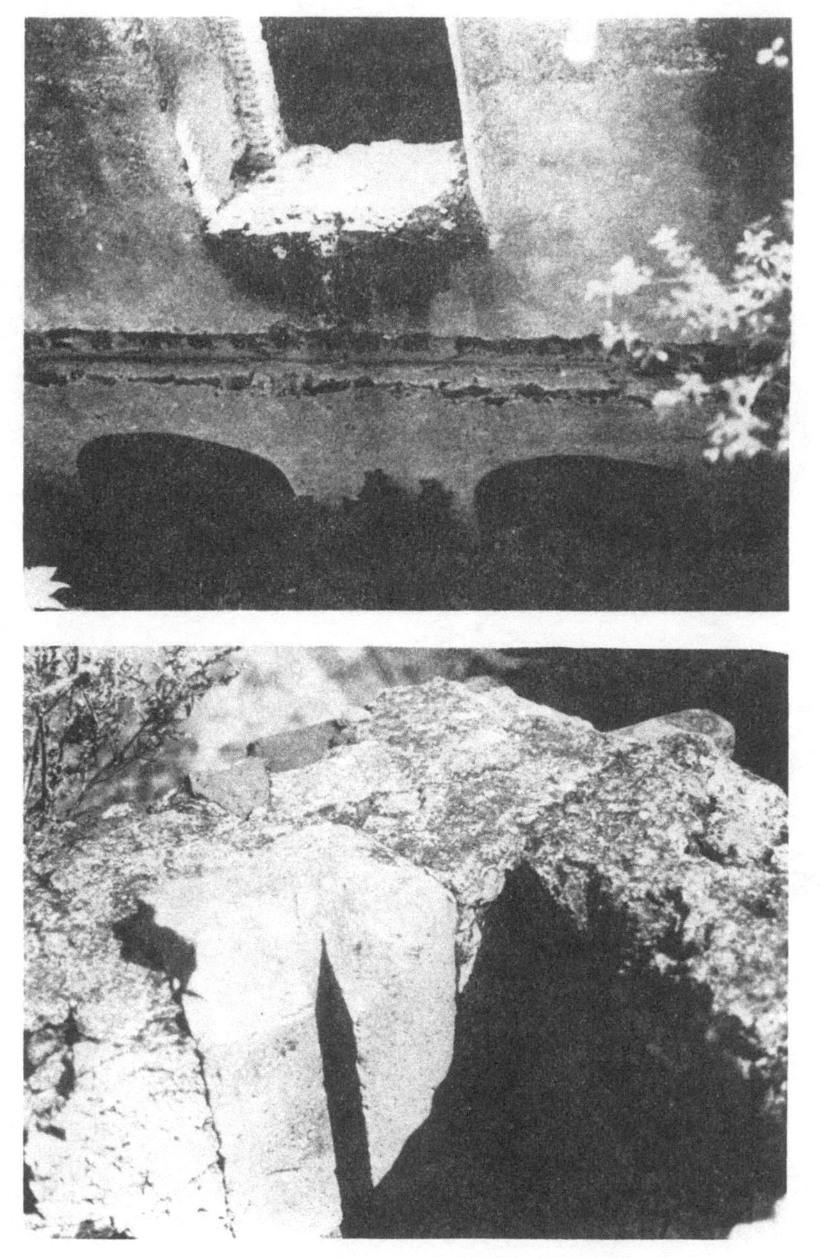

Figure 6.10 Headgates of Hacienda Espejo in 1981

Figure 6.10 View of Hacienda Espejo and irrigation features in 1981

The Hacienda Espejo itself embraces a small oval valley that lies just above the valley of Querétaro. The builder of the dam, Bernardino Primo y Jordán, intended to irrigate both this small valley and larger properties closer to the valley floor. Upon his death, he divided the upper and lower properties between two heirs; the first received the flour mill and a right to one-third of the water to irrigate six caballerías; the second received a right to two-thirds of the water to irrigate twelve caballerías. But the capacity of the dam exceeded Primo y Jordan's expectations; in most years, there was more water than needed to irrigate the eighteen caballerías. The owner of the lower properties built storage basins to receive the excess water and customarily sold part of it to other haciendas such as the Hacienda Balvaneda, located about twelve kilometers from the tunnel, which stored the water in other storage basins.[58] The irrigation canals were simple ditches; the fact that the water could be conducted such a distance can be attributed to the volume of the flow and the clayey texture of the soil.

Not surprisingly, the sale of excess water by the lower property owner led him into conflict with the upper property owner. After vigorous litigation, the parties adopted procedures that permitted a careful hour-by-hour accounting of the use of the water. The locked door to the tunnel headgate and the locked irrigation gates could be operated only with two keys, one in possession of each party, as a guarantee against surreptitious use of the water. A party that used more than his one- or two-thirds share had to compensate the other. To enable the parties to plan the use of the water, the administrator of the Hacienda Espejo devised a means of estimating the quantity of water in the dam's reservoir. He placed a measuring rod twelve varas long inside the dam to measure the level of the water and kept a record of how many hours of usage were necessary to cause the water level to descend one *pulgada*. Owing to the configuration of the reservoir, a pulgada of water near the top of the rod represented many more hours of potential water usage than a pulgada near the bottom. By tabulating the values for each pulgada of the rod, the administrator was able to closely estimate the remaining hours of usage by measuring the level of the water.[59]

The irrigation system of the hacienda was a near perfect instrument for capturing the flow of the arroyo. Although it has suffered from siltation, one may still admire the intelligence and skill of its construction. A contemporary observer conservatively estimated that the dam had a value of 18,000 pesos in 1807. It unquestionably represented much of the value of the upper and lower properties.[60]

WATER MEASURES

The water measures of colonial Mexico were not true calculations of the volume of flow, that is, of the product of the cross-sectional area and the velocity ($Q = A \times V$). Since the agrimensores lacked any means of determining the velocity of flow, the measures rested solely on a calculation of the cross-sectional area. The basic unit was

[58] Haciendas Salitre and Balvaneda "en el paraje llamado Romeral." AGN-T, Vol. 1353, Exp. 1, Cuad. 1, ff. 45, 69-72, 100v-25; Vol. 1354, Exp. 1, f. 23.

[59] AGN-T, Vol. 1353, Exp. 1, Cuad. 3, f. 5.

[60] Ibid, Cuad. 1, ff. 66v, 134.

always a square vara, commonly known as a buey of water.[61] Other units were derived from this standard. A tercia or a *quartel* of water, units frequently encountered in the seventeenth century and occasionally thereafter, represented one-third or one-fourth of a vara squared.[62] Smaller measures, such as the surco, naranja, and real that appear in the seventeenth century judicial allocation of Querétaro were fractions of one-fourth of a vara squared.[63] Early in the course of the eighteenth century, one finds widespread acceptance of a system of water measures consisting of five units--the buey, surco, naranja, real, and paja.[64] Each was assigned uniform proportions: there were three surcos in one-fourth of a vara squared and forty-eight in a buey; three naranjas in a surco; eight reales in a naranja, and eighteen pajas in a real. Other units, such as a *dedo*, though occasionally found, fell into disuse.[65]

The general acceptance of these five units of water measurement can be traced to the influence of José Sáenz de Escobar. Active around 1697-1722, Sáenz enjoyed great personal prestige. He was a fiscal and later oidor of the audiencia, and, as late as 1761, Francisco Javier Gamboa quoted his writing with deference.[66] Among other works, he wrote the *Libro de Ordenanzas y Medidas de Tierras y Aguas* which contains a detailed discussion of water measures. Although it was never printed, various copies existed and were widely read. Sáenz' influence is plainly visible in later texts which cite him,[67]

---

[61]The buey was sometimes called a *toro*. It was normally the largest unit of measurement, although a still larger measure, the *marco*, is mentioned in one case. AGN-T, Vol. 1114, Exp. 1, ff. 24v-26v; Vol. 1169, Exp. 1, Cuad. 5, f. 1.

[62]E.g., AHML, caja 1646-48, Exp. 20; AGN-T, Vol. 111, 2a pte, Exp. 2, ff. 18, 37; Vol. 185, Exp. 1, ff. 15-19; Vol. 187, Exp. 2, f. 109; Vol. 351, Exp. 1, Cuad. 2, f. 8v; Vol. 383, Exp. 4, f. 71; Vol. 2959, Exp. 141, f. 15v.

[63]AGN-T, Vol. 2648, Exp. 2 (numerous references).

[64]Besides other sources cited here, see Bribiesca Castrejón, Vol. 13(1), p. 84; AGN-T, Vol. 674, Exp. 3, Cuad. 3, f. 10; Vol. 822, Exp. 2, Cuad. 1, f. 108; Vol. 1168, Exp. 3, f. 62v; Vol. 2074, Exp. 2, f. 90v; Vol. 2716, Exp. 3, f. 17. For data regarding the diversity of water measures in Spain, see Llaurado, I, pp. 113-16. In colonial Argentina, one finds a quite different system of water measures. Wauters, passim.

[65]I have found other small measures in only two repartimientos. AGN-T, Vol. 1888, Exp. 1, Cuad. 8, f. 28v; León, II, p. 28. Reales were sometimes called *limones*. See AGN-T, Vol. 1169, Exp. 1, Cuad. 5, f. 85.

[66]Gamboa, p. 241. Sáenz served as fiscal in 1722. AGN-M, Vol. 70, f. 109. Medina identifies printed briefs that he wrote between 1697 and 1713. Medina, III, pp. 302, 359, 429, 509. An eighteenth century copy of excerpts from his writings identifies him as an oidor of the audiencia. *Ordenanzas Vigentes* (mss). His treatise discussed herein is the *Libro de las Ordenanzas* (mss) more fully identified in the bibliography.

[67]Lasso de la Vega, footnotes 9, 41, 47; AGN-T, Vol. 988, Exp. 1, Cuad. 4, f. 92v; Vol. 1362, Exp. 1, Cuad. 4, f. 22; Vol. 1888, Exp. 1, Cuad. 8, f. 53v (see drawing).

reproduce his tables and drawings,[68] or use the very small unit of linear measurement, the *grano*, that he recommended.[69]

Some eighteenth century texts relate the units of water measurement to the linear measure, the vara.[70] A surco, for example, was said to be one *sesma* (one-sixth vara) high and an *ochava* (one-eighth vara) wide. One early eighteenth century manuscript of this sort seems to betray the author's limited ability to convert one unit to another.[71] But Sáenz' de Escobar clearly saw that the units should be understood in terms of area, and he discusses the means of reducing the units to different spherical and rectangular shapes. A text on water measurement written by Lasso de la Vega in 1761 and three late colonial descriptions of water measurement in the Archivo General de la Nación share this appreciation of the units as expressions of area. In the table below, I have set forth the values of the five basic units in square pulgadas, granos, and centimeters.[72] The few sources referring to other small units of water measurement, the dedo, *sesta*, and *quinta*, suggest conflicting values.[73]

Units of Water Measurement

	Pulgadas^2	Granos^2	Centimeters^2
Buey	1296	36864	7022.44
Surco	27	768	144.30
Naranja	9	256	48.76
Real	1.125	32	6.09
Paja	.0625	1.77	.34

Occasional discrepancies can be safely explained by difficulties in the conversion of units rather than any lack of acceptance of the values themselves. Pulgadas and dedos cannot be readily converted one into another. The real equaled thirty-two granos squared; its

[68]Galván Rivera, pp. 136-37; *Tablas para la Instrucción de las Medidas de Tierras y Aguas*, passim.

[69]AGN-T, Vol. 2055, Exp. 1, Cuad. 2, ff. 27-8.

[70]E.g., Lasso de la Vega, p. 7; AGN-M, Vol. 70, f. 107.

[71]*Forma Muy Util y Provechosa para Medir Aguas* (mss), f. 31v.

[72]I assume 192 granos = one vara, and one vara = .838 meters. See Sáenz, *Libro de las Ordenanzas* (mss), p. 74; Carrera Stampa, p. 13. Carrerra Stampa's article contains an apparent typographical error on p. 23. A naranja is said to be one-third surco but to have a value of two pulgadas squared as compared to twenty-seven pulgadas squared of a surco. The correct value is nine pulgadas squared.

[73]León, II, p. 28; Orozco y Berra, p. 211; Lasso de la Vega, p. 7; AGN-T, Vol. 1888, Exp. 1, Cuad. 8, ff. 28v-31v; AGN-M, Vol. 35, f. 186.

common description as a square pulgada is a mere approximation.[74] Sáenz encountered other difficulties because of his reluctance to use the decimal system. He took the value of π to be 22/7 and, to convert a rectangle to a circular area, he first calculated the area of the rectangle and then calculated the diameter of the circle according to the formula $D = \sqrt{14A/11}$, which can be derived from the equation $A = \pi r^2$ if 22/7 is substituted for π.[75] This, of course, was a very awkward method when applied to small fractional areas and caused Sáenz to err in the calculation of the diameter of a *paja*.[76]

The three accounts of water measurements in the Archivo General de la Nación follow a similar pattern. The surveyor appeared with a measuring rod especially calibrated for the task (*vara hydromensura* or *vara sellada y graduada*). Two of the surveyors calculated the flow in certain wooden or masonry conduits having geometrical shapes, which facilitated the measurement of the cross-sectional area, but all faced the task of calculating the flow in ditches of irregular dimensions. One surveyor clearly explained his method; he measured the width of the ditch and the depth near both sides, estimated the area of a rectangle equal to the cross-section of the flow, and then calculated the units of water on the basis of this estimate.[77] Despite the inevitable inaccuracy of such an estimate, the surveyor was at an advantage in measuring an irrigation ditch. It was part of the practical art of irrigation to construct ditches so that the flow was slow enough to avoid erosion and fast enough to avoid siltation. The velocity of flow in a wooden or masonry conduit or a natural water course might vary more widely. In the case of natural streams, Sáenz de Escobar and Lasso de la Vega could only counsel surveyors to make their measurements in a relatively placid stretch of water.[78]

A distinct problem was faced in the division of water in irrigation boxes. The size of the orifice discharging water from the irrigation box was frequently calibrated to equal the cross-sectional area of the quantity of water to which a farm was entitled. A farm entitled to three surcos thus would receive water through an orifice one-fourth of a vara squared in size. But the volume of water flowing from an orifice will actually vary with the head of water above the orifice itself. The sizing of the orifice alone will have little meaning unless there is some means of reducing variations in the head of water and standardizing the shape of the box.[79] For example, a series of sugar haciendas drew water from the River Amosinac in Morelos. In an effort to achieve a proportionate division of water, a judge required each hacienda to receive water from an irrigation box having an orifice of a designated size, but he made no attempt to regulate the flow of water to the boxes. The variations in the head of water in the various boxes must have rendered the allocation nearly meaningless.[80]

---

[74]See Carrera Stampa, p. 23.

[75]Sáenz, *Libro de las Ordenanzas* (mss), ff. 43-43v. Also, Lasso de la Vega, p. 19.

[76]Sáenz, *Libro de las Ordenanzas* (mss), f. 75v. Lasso de la Vega similarly erred in the calculation of the area of a paja. Lasso de la Vega, p. 7. Galván Rivera gives the correct value. Galván Rivera, p. 134.

[77]AGN-T, Vol. 988, Exp. 1, Cuad. 4, ff. 88-90; Vol. 1888, Exp. 1, Cuad. 8, ff. 25v-31v; Vol. 2055, Exp. 1, Cuad. 2, ff. 27-28.

[78]Sáenz, *Libro de las Ordenanzas* (mss), f. 8; Lasso de la Vega, p. 13.

[79]Daugherty and Franzini, pp. 88, 382.

[80]AGN-T, Vol. 2055, Exp. 2, Cuad. 2, ff. 27v-34 and drawing, f. 48.

Some colonial experts were, however, aware of the effect of variations in the head of water on the velocity of flow through an orifice. Sáenz de Escobar offered three practical recommendations to reduce these variations. First, the orifices should be placed at the bottom of the box between parallel lines to assure that each orifice was at the same level. If a box contained orifices of one, four, and eight reales, the orifice should consist of rectangles on the same plane of the following shapes:

	height	width
One real	16 granos	2 granos
Four reales	16 granos	8 granos
Eight reales	16 granos	16 granos

Second, the boxes should be absolutely level to prevent the pooling of water above the orifice.[81] Third, the irrigation box should be five or six varas in size.[82] A smaller box might operate as a sort of surge chamber, resulting in fluctuations in the head of water above the orifice.

The very meticulous drawing of an irrigation box, shown in Figure 6.11, reveals a similar effort to reduce variations in the head of water above the orifice. As recommended by Sáenz, the orifices are placed between parallel lines. Moreover, the water was introduced into the box at a level just above the level of the discharge orifice. If the flow were increased unduly, there might be spillage or ponding around the box, but the increase in the head of water above the orifices themselves would be minimized.

Sáenz de Escobar understood that the volume of flow was actually a function of both area and velocity. He estimated the liquid volume that would pass in a minute through a paja or a circular orifice one pulgada in diameter, but offered no explanation of his method of arriving at these estimates.[83] Lasso de la Vega and a third colonial author, Alexandro de la Santa Cruz Talaban, who wrote a treatise touching on many technical subjects in 1778, taught what they believed to be an accurate, scientific method of determining the volume of flow.[84] Their method rested on the faulty premise that the velocity of flow is proportional to the depth of the stream of water and involved the determination of a point of "average velocity" in the center of a stream. The notion evidently derived from the hypothesis of Benedetto Castelli, a student of Galileo, that the velocity of flow varied with the depth of a stream of water. Though later repudiated by Castelli himself, the concept gained considerable acceptance among Italian engineers and scientists in the seventeenth century. Lasso de la Vega and (probably also) Santa Cruz Talaban gained their knowledge of hydraulics from Tomás Tosca, a Spanish mathematician, who published his major work in the mid-eighteenth century.[85] Tosca was a man of real accomplishments, but, in the field of hydraulics, he was unaware of the work of his contemporaries and drew heavily on Italian writers of the previous century. A Mexican, Alzate y Ramírez, was better informed and cited the leading contemporary

---

[81]Sáenz, *Libro de las Ordenanzas* (mss), ff. 79v-82.

[82]AGN-M, Vol. 70, f. 109.

[83]Sáenz, *Ordenanza Vigentes* (mss).

[84]Lasso de la Vega, pp. 16-19; Santa Cruz Talaban (mss), f. 106 et seq.

[85]Cuevas Aguirre y Espinosa also quotes an erroneous proposition of Tosca. Cuevas Aguirre y Espinosa, p. 33. On Tosca, see Moreno, p. 96, and Marco Cuellar, passim.

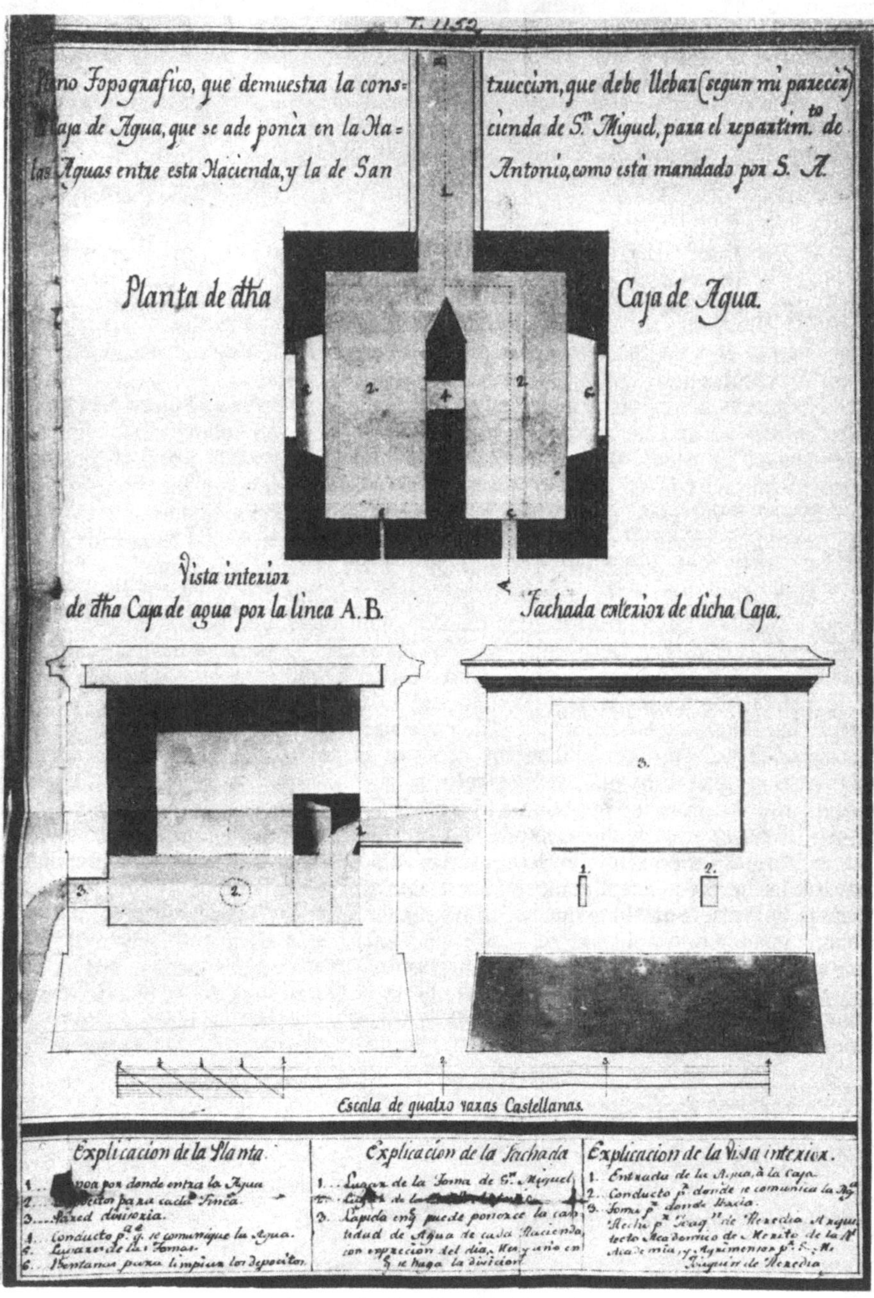

Figure 6.11 Irrigation box
(Source: AGN-T, Vol. 1152, exp.1)

figures in hydraulics.[86] But we can see that Lasso de la Vega and Santa Cruz Talaban, who accepted the authority of Tosca, were living in the province of a province, a step removed from Spain which was itself not fully in touch with new scientific developments in Europe. Moreover, their discussion of the velocity of flow reveals a significant fact: they knew nothing of the Pitot tube, the first practical device to measure the velocity of flow, which had been invented in 1732 and was then widely used in Europe.[87]

The colonial water measures died out slowly. Alzate y Ramírez ignored them completely and made an estimate of the flow of the spring at Chapultepec in the scientifically acceptable terms of cubic pulgadas per hour.[88] But Orozco y Berra and Galván Rivera sought in the nineteenth century to assign the measures precise values of liquid volume per unit of time. Their example has unfortunately been followed by several competent modern historians.[89] There may, on occasion, be some justification for accepting a colonial surveyor's estimate of the cross-sectional area of flow and making a rough estimate of velocity, but one is unlikely ever to encounter a case where the factors actually governing velocity of flow can be estimated with much accuracy.

LEVELING FOR IRRIGATION

The technology of irrigation is so intimately related to the techniques of determining a level line that the medieval irrigation specialists of eastern Spain were called *livelladores* or levelers.[90] There is much evidence that many of the architects and agrimensores of colonial Mexico were skilled in these techniques.[91] For irrigation canals, they generally favored slopes between 2.5:1000 and 5:1000, although there was a diversity of opinion on the subject[92] and a lesser slope might be needed in a masonry conduit. The architects of the Querétaro aqueduct calculated that it would descend fourteen varas over a distance of about 8,500 varas.[93] The fact that the aqueduct was completely successful suggests that their calculations were tolerably accurate. It is noteworthy in any event that the surveyors had confidence in their ability to make such a

[86]Alzate y Ramírez, III, p. 413.

[87]Rouse and Ince, p. 115.

[88]Alzate y Ramírez, II, p. 19.

[89]Galván Rivera, p. 135; Orozco y Berra, p. 211; Carrera Stampa, p. 23; Prem, p. 65; Santa Cruz F., pp. 247-51.

[90]Glick (1968), passim.

[91]E.g., *Estado General de las Fundaciones*, II, pp. 48, 68, 119. (The engineer D. Agustín López measured level lines quickly and accurately, if we can believe the account.) *Relaciones del Desagüe del Valle*, pp. 34, 52-72, 82-83, 225. Consider also the design of the Hacienda Espejo.

[92]Sáenz, *Libro de Ordenanzas* (mss), f. 6; Moreno (1977), p. 336; Olson and Eddy, p. 107.

[93]AGN-M, Vol. 70, f. 105v.

fine calculation and that others trusted in its accuracy. How did the architects measure this gentle slope through a long, winding, and uneven course? We may once again turn to Sáenz de Escobar's treatise on water measurement, which contains a chapter on leveling instruments.[94] Although the treatise is undated, Sáenz was active as a legal advisor to the audiencia when the aqueduct was built and may well have written around that time.

Sáenz regarded a simple plumb bob level as being one of the essential instruments of a surveyor. In his treatise, he suggests that the level could be fitted with a quadrant marked with degrees and gives a drawing of such an instrument (Figure 6.12a). The level shares a generic similarity with the drawing of a simple A-frame level appearing in a contemporary surveying treatise (Figure 6.12b).[95] But Sáenz observes that this simple form of level may be inadequate for measurement of a level line over a long distance, and he enters into a learned discussion of other leveling instruments. He resorts, first, to the authority of Vitruvius who mentioned the use of three instruments--the *dioptera, libra aquaria*, and *chorobate*. All three were water levels of the type first developed by the Greeks. Despite Vitruvius' interest in these instruments, they never enjoyed much acceptance in Roman times and fell out of use in the Middle Ages. A few savants of the sixteenth century once again proposed the use of water levels, but it was not until the seventeenth century that instrument makers began actively to revive and refine the ancient Greek technology.[96] Sáenz knew of water levels chiefly through a work of Gaspar Schott, a German Jesuit, who described an array of water levels in his *Pantometrum Kircherianum* published in 1660. Schott was a devoted disciple of the Jesuit scholar, Athanasius Kircher (1602-1680), who distinguished himself as a disseminator of new scientific knowledge. In his Roman residence, Kircher assembled an extensive collection of surveying instruments, which Schott discussed one by one.[97] Sáenz' discussion of leveling instruments is largely a summary of material found in Schott's treatise, but his drawing of a chorobate, a large, cumbersome table fitted with plumb bobs and a water trough that could be propped up to a level position with difficulty, appears to be a reconstruction of the chorobate described by Vitruvius.[98]

Were the water levels Sáenz describes actually found in colonial Mexico? His language is too ambiguous to permit a confident answer, but his discussion of leveling techniques should be regarded primarily as a display of erudition rather than a description of techniques actually in use. For example, he proceeds from a description of instruments to a discussion of the use of triangulation in calculating a slope. The measurement of angles, in fact, could be easily avoided by the use of a measuring rod in surveying a gentle slope. But Sáenz' discussion of leveling instruments and practices illustrates well the channels by which scientific knowledge reached Mexico under the late Hapsburgs. He cites five authorities; all were clerics and four were Jesuits.[99] Moreover, his discussion is notable for the fact that he omits any mention of the use of the optical lens

---

[94]Sáenz, *Libro de Ordenanzas* (mss), f. 60v.

[95]*Forma Muy Util y Provechosa para Medir Aguas* (mss), f. 57.

[96]The following discussion of the history of leveling instruments draws from Richeson, passim; Kiely, pp. 10-37, 54-59, 129-43; Daumas, pp. 76-78; Stone, passim.

[97]Schott, pp. 269-323; Gillispie, XXI, p. 210.

[98]The drawing appears in the copy of Sáenz' treatise in the Biblioteca Nacional, Mexico City, but not in the copy in the Bancroft Library, Berkeley.

[99]Jesuits: Gaspar Schott, Athanasius Kircher, Cristoph Clavius, Nicolás Cabeo. Cistercian: Juan Caramuel y Lobkowitz.

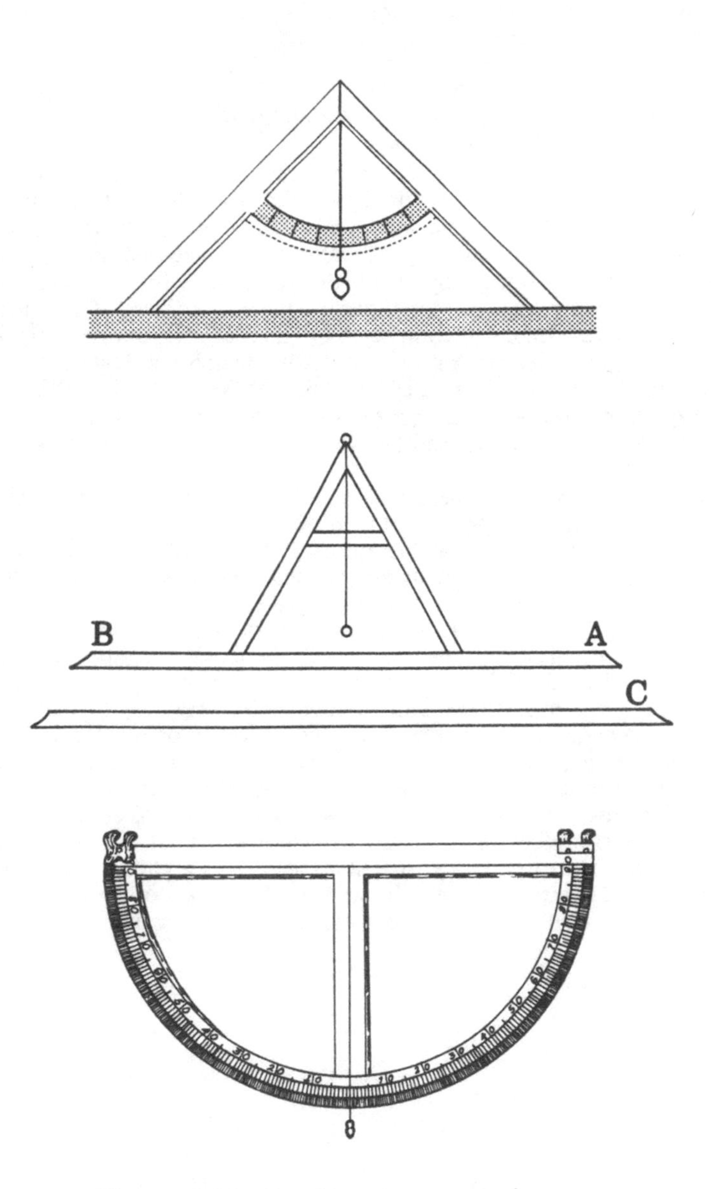

Figure 6.12 Leveling instruments
(6.12a top; 6.12b center)

and the bubble level. The telescopic lens, invented in 1608, was used by Picard to design the water system for the palace gardens of Versailles in the 1680s, and began to find other applications for surveying in the late seventeenth century. It came into general use in surveying instruments when the techniques for grinding good lenses of small size were perfected around 1760. The bubble level, first described by Thenot in a tract published in 1666, had the great advantage of compact size and portability. La Hire reported that this kind of level was "en fort usage" in 1689,[100] but, until techniques were perfected to bore a glass tube with a precise diameter in the mid-eighteenth century, the bubble level competed with water levels and plumb bob levels.

Despite Sáenz' personal erudition, his treatise reveals the technological isolation of Mexico in the early eighteenth century. It negates the possibility that the optical lens or bubble level had spread to this province of western culture and contains unambiguous evidence of only a limited literary knowledge of water levels. The traditional plumb bob level was clearly the instrument in general, if not exclusive, use. In 1761, Francisco de Gamboa described the instruments used for surveying mines. Although the demands of this sort of surveying were different from those in the field of irrigation, it is significant that the level which Gamboa depicts is a variation of the plumb bob level described by Sáenz (Figure 6.12c).[101] The levels used in the design of the Querétaro aqueduct may, however, have been more sophisticated than the simple devices shown in Figure 6.12. Plumb bob levels came in many types, and the best remained in use until the end of the eighteenth century.

The Enlightenment began to penetrate colonial Mexico in the mid-eighteenth century through the efforts first of a few Jesuit scholars and later of a more diverse group of intellectuals. The decisive event was the approval in 1775 of Gamarra's *Elementa Recentioris Philosophie* both by his religious superior and the inquisition itself. After that date, Miranda writes, "the movement spread rapidly and its interest centered more on science than philosophy."[102] One of Gamarra's censors was Joaquin Velázquez de León, a man of great technical competence, whom Humboldt describes as being as accomplished as a French academician in his astronomical observations.[103]

Velázquez de León has left us a description of the leveling instruments and methods he used in preparing a topographical map of the valley of Mexico in 1773-75. His lucid, well-organized prose is a world removed from the turgid erudition of Sáenz de Escobar. One senses an emergence from the Middle Ages to the modern era in a mere fifty years. Velázquez used two leveling instruments. Both employed an air bubble and an optical lens that were of recent manufacture in France and England. In calculating level lines, he followed the treatise of La Hire, still regarded as a classic text in the late eighteenth century.[104] He was, in short, fully abreast of the state of the art in Europe. Velázquez de León was well in advance of some of his contemporaries,[105] and his

---

[100]La Hire, p. 143.

[101]Gamboa, p. 260, plate no. 2.

[102]Miranda (1962), p. 32.

[103]Moreno (1977), p. 60.

[104]Ibid, pp. 315-18.

[105]Santa Cruz Talaban's discussion of leveling techniques consists of a belabored treatment of a false problem: the effect of the curved surface of the earth on the measurement of a level line. Santa Cruz Talaban (mss), f. 106.

instruments plainly were a rarity,[106] but he participated in a general revival of technical competence. The maps and drawings of the agrimensores in the Bajío tend to be of markedly improved quality in the latter half of the eighteenth century. The meticulous draftsmanship of such drawings as Figure 6.7 surely reflects progress in instrumentation and surveying practices.

WATERWORKS

The persons who built the waterworks of the Bajío are not often identified in the surviving documents. The construction was no doubt sometimes entrusted to masons or other craftsmen, supervised by a hacienda administrator, as was the case of the dam on the Hacienda Mendoza. The builder of the early dam serving the Valle de Santiago was described as a carpintero.[107] But architects are also mentioned both in connection with major projects and the replacement of minor facilities, such as irrigation gates. The execution of the 1654 judicial allocation at Querétaro,[108] the construction of the Querétaro aqueduct,[109] the replacement of headgates in Salvatierra,[110] and the construction of the unsuccessful dam on the Lerma near Salamanca in the early seventeenth century[111] were all the works of architects. A second profession, the agrimensores or surveyors, had perhaps more importance in the design and administration of water works. Agrimensores frequently appear as authorities on water measurements and as experts employed by parties to water litigation. Members of the profession studied the feasibility of the new canal for Salvatierra in 1646,[112] prepared the plan to exploit Lake Yuriria for the benefit of the Valle de Santiago,[113] assisted the architect in the design of the aqueduct at Querétaro, designed a system to utilize water from the swamp north of Celaya,[114] and prescribed modified headgates and water dividers for the Hacienda Espejo in 1807.[115]

---

[106]Alzate describes the measurement of a level line through use of "un vaso de cristal," a method that he conceded is inaccurate. Alzate y Ramírez (1768), III, p. 146. The instruments of Velázquez de León evoked the admiration of his peers. Cajori, pp. 58-61.

[107]Gonzáles, p. 233.

[108]AGN-T, Vol. 2648, Exp. 2, f. 65 (maestro en architectura).

[109]AGN-M, Vol. 71, f. 310; Vol. 70, f. 105v (persona inteligente en architectura).

[110]BN-AF, caja 47, doc. 1059.4.

[111]AGN-M, Vol. 35, f. 21v (maestro en architectura).

[112]AGN-T, Vol. 185, Exp. 1, Cuad 1, f. 15 (medidor de tierras y pesador de agua).

[113]AGN-A, Vol. 97, Exp. 2, f. 10 et seq.

[114]AGN-T, Vol. 2072, Exp. 1, Cuad. 8 (see explanation of drawing).

[115]AGN-T, Vol. 1353, Exp. 1, Cuad. 2, ff. 31, 38.

The architects of colonial Mexico had a distinguished tradition of urban construction that reflected a well-organized system of professional training and standards. Free of the derivative character one might expect in a provincial society, colonial architecture displayed a robust self-confidence and was often marked by good construction, a refined sensibility, and a spirit of innovation. The training and certification of architects were controlled by a professional guild. Toussaint explains:

> Whoever desired to be a master architect had to work as an apprentice for six years, under contract with a certified architect, in one of three areas: mortar work, rock cutting or drawing. In addition, the apprentice needed to read, write, count and know the principles of geometry and how to "montear, reducir, cuadrear y cubicar". The examination consisted of two parts: the first of a practical nature, consisting of work on a construction project for however long the overseers of the guild should require, and the other technical, that is, an examination in theory.[116]

The architects who were called upon to design water works in the Bajío presumably had a practical foundation for their work.[117] Less is known of the profession of agrimensores. Gamboa lamented the scarcity of professional geometricians and warned against incompetent practitioners.[118] But standards of professional competence began to emerge in the second half of the eighteenth century. The audiencia followed the practice of awarding the prestigious designation "agrimensor titulado" to experienced surveyors who passed an examination. A practicing agrimensor, who sought this title in 1771, was examined in "aritmética, geometría, idrographía."[119] Toward the end of the century the mining court similarly certified surveyors of established competence.[120] The nature of the profession is best suggested by the title of "mathematical practitioner," applied to their contemporaries in England.

* * *

Any discussion of the technology and the culture of late colonial Mexico must take into account the opinions of Alexander von Humboldt. Though sometimes unduly critical of the Mexicans who helped him so much in his research, Humboldt was always a perceptive observer. He acknowledged the excellence of Mexican architecture and the lively scientific interests of many Mexicans. But he charged that the intellectual energies

[116]Toussaint, p. 297.

[117]The aqueducts of Mexico attest to architectural competence. Romero de Terreros, passim.

[118]Gamboa, Chap. 12.

[119]AGN-M, Vol. 81, f. 54.

[120]AGN-C, Vol. 1298, Exp. s/n, f. 25 (Ojo Zarco); AGN-T, Vol. 1170, Exp. 1, f. 401 (drawing); Vol. 2072, Exp. 1, f. 16.

of the colonial elite were generally diverted to nonutilitarian ends. In particular, the practical application of technology in irrigation was neglected.

> Moreover, the owners of large farms seldom feel the need to employ engineers who know how to level the ground and who understand the principles of hydraulic construction. Rather, in Mexico, as in some other places, people have favored the arts that please the imagination over those that are needed to supply the needs of domestic life. Consequently, there are architects who judge perfectly the beauty and order of a building design; but nothing is more rare than persons capable of building machines, dikes and canals. Fortunately, a sense of necessity has stimulated national labors, and a certain sagacity, common to all mountain peoples, has offset to some degree the lack of instruction.[121]

It is not easy to put Humboldt's judgment to the test of empirical investigation. Although we know much about the intellectual life of the capital, the popular technologies of the provincial towns and countryside are difficult to discover. Certain areas of agricultural technology were unquestionably marked by a resistance to innovation and the persistence of traditional, labor intensive methods. Morin observes that the simple agricultural implements with few, if any, metal parts noted in estate inventories reflect little interest in efficient use of labor.[122] In an excellent study, Basave Kunhardt shows that haciendas characteristically manufactured and repaired their own tools in an effort to achieve relative self-sufficiency and were very slow to take advantage of improved factory-manufactured equipment and new agricultural machinery in the nineteenth century.[123] The factors responsible for this persistence of traditional methods of cultivation and harvesting--above all, cheap labor and limited markets--also affected some irrigation systems. The occasional use of crude earthen dams, for example, was favored by low cost of labor that could be mobilized to replace them. But such dams had limited application. Most waterworks bring us into a different sphere of technology characterized by significant capital investment and the activity of professional specialists.

On the whole, Humboldt seems to have given insufficient credit to the work of architects and agrimensores in the irrigation systems of colonial Mexico. Our investigation suggests a more complex and nuanced view. Although some irrigation systems used crude labor-intensive methods, many employed impressive masonry constructions designed by competent professionals, and, while the effects of provincial isolation are often evident, one finds drawings and measurements manifesting a high degree of technical skill. The picture that emerges is that of a provincial society, in only partial contact with the centers of innovation in Western Europe, but possessing a local tradition of excellent architecture and some competence in surveying that found occasional application in waterworks.

---

[121]Humboldt, II, Book 4, Chap. 9, p. 422.

[122]Morin, p. 243.

[123]Basave Kunhardt, passim. Also Bullock, p. 277.

7
Water Law

Colonial water law was a system of rights, conferred and administered by a central authority, on which was engrafted certain rules regarding possession, easements, and natural water courses taken from Spanish law. In practice, it was sometimes affected by customary irrigation practices originating in Spain. The royal decrees collected in the *Recopilación de Indias* established the basic framework of an administered system of rights, but, except for decrees regarding Indian pueblos, they contained few provisions that had a direct bearing on the resolution of water disputes. The Recopilación de Indias provided, however, that matters not governed by its provisions were to be decided according to the law of Castile.[1] Consequently, many proceedings were affected by the *Siete Partidas*, the legal code written in the late thirteenth century, and by the later compilations of Castilian law, the *Nueva Recopilación* of 1567 and the *Novísima Recopilación* of 1805. The Siete Partidas, which enjoyed a prestige somewhat comparable to the constitution of the United States, contained many provisions regarding water rights, and the later compilations included certain procedural rules of great practical importance in water litigation. At least two legal treatises also had a possible influence on water law: the gloss on the Siete Partidas of Gregorio López and the *Política Indiana* of Juan de Solórzano Pereira. Like other contemporary treatises, they are both largely devoted to the discussion of Roman law from which Spanish law evolved.

The upper class in central New Spain was in the main a legally sophisticated society.[2] Lawsuits proceeded according to rules that are bewildering to the modern researcher but were understood by contemporaries. The citation of statutory authority was fairly frequent but not always precise. In my research, I have encountered eighteen citations of the Siete Partidas, about one-half of which referred to the precise title and law. But even where laws are not expressly cited, it is often apparent, at least in the eighteenth century cases, that the lawyers were well versed in the Recopilación de Indias, the Nueva Recopilación, and the Siete Partidas. Some decisions were, indeed, the sort of

---

[1]RLI, 2:1:2. For explanation of the phrase "la de Toro," see Spota, I, p. 282-83.

[2]Malagón Barcelo, passim.

"pragmatic judgements," unaffected by formal law, that Taylor describes.[3] But at the other extreme, one occasionally encounters touches of erudition. The capable lawyer in a Querétaro water case drew from the Justinian Digest for a rule favorable to his client.[4] Sáenz de Escobar, who reviewed a petition of the farmers of the Valle de Santiago as legal advisor to the audiencia, clearly possessed much learning in the jurisprudence of his day. Among the colonial lawyers that I have encountered, Miguel Domínguez was in a class by himself. Gifted with a fine legal mind, Domínguez was to gain a kind of immortality not for his own considerable attainments, but as the husband of La Corregidora, Josefa Ortiz de Domínguez, whom he married a year after the birth of their first child. The legal analysis in his decisions would do credit to an appellate judge of the twentieth century.[5]

A difficulty in interpreting colonial law must be noted at the outset. The audiencia rarely gave an explanation of the basis for its decisions; the standard form of order consisted of only a short paragraph that recited the previous petitions and orders and stated the disposition of the case. The problem is compounded by the fact that a large portion of the water cases in the Archivo General de la Nación are mere fragments in which the outcome of the case cannot be determined. Local archives, which in some areas may contain a reserve of supplementary documents, were of almost no assistance in the Bajío. It is true that the legal advisor to the audiencia sometimes offered a reasoned recommendation of action to be taken, and a few proceedings before local or commissioned judges, such as Miguel Domínguez, contain a complete transcript and a detailed explanation of the decision. But in reviewing the many alternative arguments in water law litigation, one cannot always be sure whether a particular legal theory actually influenced the outcome of the case.

The Recopilación de Indias, though infrequently cited in water cases, clearly had a pervasive impact on the nature of the proceedings. A premise underlying much of water law is stated in Book 3, title 1, law 1.

> Por donación de la Santa Sede apostólica y otros y justos y legítimos títulos, somos señor de las Indias Occidentales, Islas y Tierra-firme del mar Oceano, descubiertos y por descubrir, y están incorporadas en nuestra real corona de Castilla.

---

[3]Taylor (1975), p. 200. In support of his contention that the water cases were little affected by formal law, Taylor states that, in the thirteen eighteenth-century cases he read, he did not find a single citation of the Recopilación de Indias. But in researching this essay, I have found three citations of the Recopilación de Indias, two citations of Roman law, ten citations of the Nueva Recopilación, three discussions of legal commentaries, and eighteen citations of the Siete Partidas. The learning of the two colonial writers on water rights, Sáenz de Escobar and Lasso de la Vega, was shared by many of their contemporaries.

[4]AGN-T, Vol. 1354, Exp. 1, ff. 100v, 165.

[5]AGN-T, Vol. 1362, Exp. 1, Cuad. 3, f. 86. Also AGN-T, Vol. 1168, Exp. 3, f. 60v; Vol. 1353, Exp. 1, f. 39v.

Relying on the papal bull *Inter Caetera* [6] and the feudal rights of a sovereign over conquered territory, the Crown asserted an unqualified dominion over the Indies.[7] The land, together with the water resources, belonged to the royal patrimony: it was property of the Crown. Thus, Solórzano wrote:

> Nos es digno de menor consideración otro derecho que compete y está reservada a los Reyes y Soberanos Señorios por razón de la suprema potestad de sus Reynas y Señorios, conviene a saber, el de las tierras compos, montes, pastos, ríos y aguas públicas de todo ellos.[8]

Exercising this royal supremacy over the Indias, the Crown decreed that the waters were for the common use of the population.

> Nos, hemos ordenado, que los pastos, montes y aguas sean comunes en los Indias...Mandamos que el uso de todas los pastos, montes y aguas de las Provincias de las Indias, sea común a todos los vecinos de ellas que ahora son y después fueren para que los puedan gozar libremente y hacer junto a cualquier buhío sus cabañas, traer allí los ganados, juntos o apartados, como quisieren...[9]

But this right of common use, which had several antecendents in the Siete Partidas,[10] was understood to apply only to the use of streams for drinking, fishing, and domestic purposes; it did not permit the free diversion of water for irrigation and, arguably, forbade unauthorized private appropriations.[11] The author of a treatise on water rights, Lasso de la Vega, explained:

> No se presuma haber de ser públicos y comunes en cuanto a su conducción: pero si en cuanto a su uso personal y doméstico.[12]

The right of common use figured chiefly in the few cases where there was a conflict between domestic and other uses. For example, when an obraje in Querétaro sought to

---

[6]The papal donation to which the law refers was *Inter Caetera* of May 4, 1593. See Florescano (1976), p. 23.

[7]RLI, 4:12:14.

[8]Solórzano Pereira, V, Book 6, Chap. 12, p. 37. To the same effect: Lasso de la Vega, p. 1; Sáenz, *Libro de las Ordenanzas* (mss), f. 73v. The Bourbon monarchs asserted a somewhat analogous dominion of the rivers of Spain in the eighteenth century. See NovR, 7:11:24 (20) and 27 (48).

[9]RLI, 4:17:5.

[10]SP, 3:28:3, 6, 31. Also RLI, 4:17:5, 7.

[11]The argument is made in AGN-T, Vol. 1169, Cuad. 5, ff. 68, and 162 based on the Siete Partidas.

[12]Lasso de la Vega, p. 3.

appropriate a spring that was a vital source of drinking water, the residents of the town objected that "Los ojos de agua de la Cañada son comunes al cuerpo todo de vecinos así indios como españoles."[13]

A logical corollary of the Crown's assertion of dominion over the Indies was that only the king, or his designated representatives, had the ultimate right to grant individuals the title to land and water. Several provisions of the Recopilación de Indias served to delegate the exercise of this right, subject to certain norms, to the viceroys of the Indies.[14] For example, Book 4, title 12, law 5 stated:

> Habiéndose de repartir las tierras, aguas y abrevaderos y pastos entre los que fueren a poblar, los vireyes o gobernadores que de Nos tuvieren facultad, hagan el repartimiento, con parecer de los cabildos de la ciudades, o villas, teniendo consideración a que los regidores sean preferidos.

Viceregal grants (mercedes) of water rights, separate from land grants, were quite infrequent in most regions,[15] but land grants, which included a grant of water rights, were of great importance in establishing an orderly regime of water rights in Celaya, the Valle de Santiago, and Salvatierra.[16] The concession of water rights followed three forms. The early land grants in Celaya and a few in Salvatierra included "caballerías de riego." These concessions of irrigable land were clearly intended, and in fact were construed, to confer water rights. Later grants in Celaya and most of the grants in the Valle de Santiago, including grants in Salvatierra, followed a form that was probably

[13]AGN-T, Vol. 2765, Exp. 11, f. 11.

[14]See also RLI, 4:7:14 and 4:12:8; Florescano (1976), p. 23 et seq.

[15]Prem, p. 59. Taylor found that of over 4,000 grants in volumes 1-31 of the Ramo de Mercedes, only 101, or 2.25%, concerned rights to water separate from land. Taylor (1975), p. 197. My own review of the index of mercedes indicates that the figure is accurate.

[16]Taylor greatly underestimates the importance of land and water grants. He states that he made a "careful search" through volumes 1-31 of the Ramo de Mercedes, and he reports, "For the period 1542-1616, I have located twenty-one examples of mercedes combining land and water. The earliest example of a land and water grant in the Ramo de Mercedes dates from 1584." In my own research for this essay, I have found sixty-five mercedes combining land and water rights in volumes 1-31, of which only seven were noted by Taylor. Twenty-nine were granted in 1584 or earlier. Noted by Taylor: AGN-M, Vol. 26, ff. 67v, 160v, 161; Vol. 28, f. 64; Vol. 30, f. 100; Vol. 31, ff. 100v, 160v(1). Other such grants in 1584 and earlier: AGN-M, Vol. 10, ff. 16, 16v, 17, 69, 182v, 183, 202v, 210v, 213v, 222; Vol. 11, ff. 95v, 287v; Vol. 12, ff. 17v, 19, 27v, 45, 46, 46v, 63v, 150, 150v; Vol. 13, ff. 28v, 62v, 64v, 97, 102, 109v, 110v. Others after 1584: Vol. 14, f. 165v; Vol. 16, 45v; Vol. 18, f. 216; Vol. 19, ff 105, 186; Vol. 23, f. 84v; Vol. 24, f. 174; Vol. 26, ff. 19, 67, 67v, 139, 145, 148v, 157v, 160v, 161; Vol. 28, f. 64; Vol 30, ff. 100, 100v; Vol. 31, ff. 100v, 160v(1), 238, 239v, 240, 248, 248v, 260v, 188v. The inclusion of water rights in land grants was common enough that Sáenz de Escobar included a section on the interpretation of the phrase, "con el agua necessario para su riego," in his treatise, written for the benefit of local judges. They evidently could be expected to encounter the clause repeatedly in the performance of their functions. Sáenz, *Libro de las Ordenanzas* (mss), f. 73v.

most common throughout Mexico:[17] following a description of the land, one finds a phrase such as "con el agua necessario de Río Grande por su riego."

Land grants were subject to certain standard conditions based largely on royal decrees later collected in the Recopilación de Indias. A grant of land and water rights in Salvatierra in 1585 contained the following conditions:[18] (1) the grant was made "sin perjuicio de derecho de su Mag. ni de otro tercero"; (2) the land was to be cultivated within one year; (3) the fields were to be left for common pasture after harvest; (4) if a town were established on the site, the land could be appropriated on payment of its value; (5) the grantee could not sell the land within four years; and (6) neither she nor her successors could sell it at any time to a church, monastery, or "persona eclesiástica." If its conditions were met, such a grant provided a right to water which, though generally ambiguous, was always respected by judicial authorities in our sample cases.[19] It is true that later assertions of royal prerogatives affirmed the Crown's right to revoke or modify grants under certain circumstances,[20] but these legal refinements never affected land or water grants in the Bajío.

The *sin perjuicio de tercero* clause, included also in some judicial orders,[21] had much potential significance. In later grants, the clause was expanded to refer to "indios o otro tercero."[22] The fact that all grants and many adjudications of water rights could be enjoyed only in a manner "sin perjuicio de tercero" introduced an element of flexibility into the administration of water rights and allowed individual grants to be construed in the common interest. The clause had a vital impact on the administration of water rights in the north of Mexico,[23] but, curiously, it was rarely mentioned and never explicitly invoked as a basis for decision in the water cases of the Bajío.

According to the theory of royal absolutism, land or water that was not subject to a grant (*merced*) remained part of the royal patrimony. In the adjudications of water rights, there was considerable tension between this principle and institutional realities. Early mercedes of water rights were concentrated in certain regions. They were abundant in most irrigated portions of the Bajío but uncommon in Querétaro. Taylor has found little evidence of such mercedes in Tulancingo or the Valley of Oaxaca.[24] Moreover, landowners who built new irrigation systems in the Bajío during the eighteenth century usually did not secure formal mercedes of water rights. The principle that the flow of streams, not expressly granted to individuals, belonged to the Crown as its royal property, could not be applied rigorously. As we will see, judicial decrees became a basis of water rights, and landowners lacking any formal source of title actually had considerable success in defending their use of water through various legal strategies. But the principle

[17]Taylor (1975), p. 197; Sáenz, *Libro de las Ordenanzas* (mss), f. 81v.

[18]AGN-M, Vol. 12, f. 150.

[19]NovR, 3:5:1.

[20]RLI, 4:12:20; Solórzano Perera, V, Book 6, Chap. 12, p. 39.

[21]It was required to be included in grants by RLI, 4:12:4. Judicial orders with the clause: AGN-T, Vol. 307, Exp. 2, f. 110; Vol. 764, Exp. 3, f. 16.

[22]The specific mention of Indians was in compliance with RLI, 4:12:9 (promulgated June 11, 1594).

[23]Michael C. Meyer (personal communication).

[24]Taylor (1975), p. 199; Taylor (1972), passim.

still retained considerable vitality in colonial law. Lasso de la Vega characteristically wrote:

> Ciñéndome precisamente a la de las aguas, para norte y fundamento de todo esta Reglamento, hallo que de la misma suerte son de Regio Patrimonio que las demas bienes, que como tales estan anexas e incorporadas en su Real Corona, veniendo de aqui la denominación de Realengos, en tanto grado, que para haber de poseerlas, es menester que los particulares poseedores aleguen y prueben les han sido concedidas por especial merced de los mismos Reyes y Católicos Señores o en su nombre.[25]

In water cases, one sometimes encounters the objection that a party lacked a merced of water rights,[26] that a stream appropriated by an individual was part of the royal patrimony,[27] or that there could be no rights of prescription on public or Crown property.[28] By denouncing another's lack of title to water, a landowner could secure a grant of the water for himself.[29] The ownership of land bordering on a stream conferred no right whatever to water. Landowners possessing mercedes of water rights could obtain an order enjoining other riparian landowners from appropriating water for irrigation.[30]

The respect accorded to formal grants of water rights is illustrated by a case in San Miguel El Grande.[31] Antonio Lanzagorta owned two haciendas on the Laja, the Cieneguilla and the Petaca. In 1753, he obtained a viceregal grant of three surcos of water for the Hacienda Cieneguilla and proceeded to build a dam that captured the entire flow of the river. He later tried unsuccessfully to secure a grant of water rights for the Hacienda Petaca. Undeterred, his widow constructed a second dam on the river for this hacienda. Aroused by her action, the farmers of Celaya jointly challenged the use of both dams in 1788. The commissioned judge ordered that the dam of the Hacienda Petaca be removed and the headgate of the dam of the Hacienda Cieneguilla be modified so as to receive only three surcos. An even more striking example of the significance of formal grants of water rights was uncovered by Prem in his study of the Cotzala Valley. With the exception of one hacienda, all the irrigated haciendas in the Cotzala Valley in 1673 either possessed grants of water rights or agreements to share water with a hacienda having such a grant. A judicial allocation of water more precisely defined the rights of

---

[25]Lasso de la Vega, p. 1. Also Solórzano Pereira, V, Book 6, Chap. 12, p. 38.

[26]E.g., AGN-T, Vol. 586, Exp. 8, f. 16v; Vol. 1168, Exp. 3, f. 3; Vol. 3618, Exp. 1, Cuad. 2, f. 22v.

[27]E.g., AGN-T, Vol. 533, Exp. 2, f. 42v; Vol. 2782, Exp. 3, f. 24; AHML, caja 1649-52, Exp. 13b.

[28]AGN-T, Vol. 1390, Exp. 3, f. 9v; Vol. 3618, Exp. 1, Cuad. 2, ff. 28-9.

[29]AGN-M, Vol. 70, f. 14v.

[30]AGN-M, Vol. 29, f. 122. Also AGN-T, Vol. 544, Exp. 3, f. 19.

[31]AGN-T, Vol. 1168, Exp. 3.

the haciendas possessing formally documented rights but denied the single hacienda lacking such rights permission to irrigate.[32]

Landowners with defective titles to land or water enjoyed an opportunity to legitimize their possession in the land and water compositions of the seventeenth century. The proceedings had their origin in two royal decrees in 1591: the first ordered that all land held without proper title be restored to the Crown; the second offered to make a"cómoda composición" with landowners--in exchange for the payment of money, the Crown would issue new titles. The decrees were not effectively enforced until the period 1642-48; they were then administered as simple tax assessments under threat of dispossession of landowners. Celaya, for example, was assessed 20,000 pesos and a judge was dispatched to raise this amount. The sum was prorated among all haciendas in the jurisdiction. After prolonged negotiations between the cabildo and the audiencia, a payment was made in satisfaction of the landowners' obligations in the jurisdiction.[33] In theory, payment of the assessment operated to clear title to water as well as land.[34] The commissioned judge was commonly known as "juez privativo de tierres y aguas," and the standard decree acknowledging payment of the assessment confirmed title to water as well as land. In the case of Celaya, it confirmed title "asimismo de las aguas en cuyo uso y posesion se hayan."[35] The point was not lost on colonial lawyers who argued the effect of the compositions in several water cases.[36] But the argument always appears as one of several alternative theories; it is not clear whether it had a major impact on the actual evolution of water rights.

Indian communities claiming rights to water enjoyed a favored legal status. An early royal decree ordered that land grants should not dispossess Indian communities: "y a los indios se les dejen sus tierras, heredades y pastos, de forma que no les falte lo necessario."[37] Pre-conquest possession was always respected as a source of title.[38] A later sixteenth-century decree concerning Indian pueblos required: "confirmándoles en lo que ahora tienen y dándoles de nuevo lo necesario."[39] Other sixteenth century decrees, having specific reference to water rights, ordered that the Indians should be given the waters that they needed[40] and that indigenous systems of irrigation should be preserved.[41] The laws directing the establishment of new Indian pueblos sought

---

[32]Prem, pp. 61-64.

[33]AGN-M, Vol. 44, ff 1-23; Vol. 45, f. 70; AGN-T, Vol. 185, Exp. 1, Cuad. 2, f. 36; Velasco y Mendoza, I, p. 352; Zamarroni, I, p. 100.

[34]For a general discussion of land compositions, see Chevalier (1952), pp. 348-63.

[35]Zamarroni, I, p. 103.

[36]AGN-T, Vol. 192, Exp. 1, f. 50; Vol. 307, Exp. 2, f. 2v; Vol. 383, Exp. 4, f. 1v; Vol. 1169, Cuad. 5, f. 67v; Vol. 1175, Exp. 4, f. 60.

[37]RLI, 4:12:5.

[38]López Sarrelangue, p. 131.

[39]RLI, 4:12:14.

[40]RLI, 4:2:63.

[41]RLI, 4:17:11.

paradoxically to preserve the property rights that the Indians previously enjoyed.[42] A decree in 1573 provided:

> Los sitios que se han de formar pueblos y reducciones, tengan comodidad de aguas, tierras y montes, entradas y salidas, y labranzas y un exido de una legua de largo...[43]

The decrees in the seventeenth century provided that land compositions should not legitimize land unlawfully taken from Indians.[44] When the Indian population had sunk to its lowest point, royal decrees became most demanding. A 1646 decree ordered that special consideration be given to Indian communities in cases of competing land claims.[45] A second decree issued the same year, explicitly referring to water rights, ordered that the communities be given more than they needed for their sustenance.

> Ordenamos que la venta, beneficio y composición de tierras se hagan con tal atención, que a los indios se les dejen con sobre todos las que les pertenecieren asi en parrticular como por comunidades, y las aguas y riegos; y las tierras en que hubieren hechas acequias u otro cualquier beneficio...[46]

The policy was reaffirmed in the eighteenth century.[47]

This constellation of laws, which could be much elaborated and extends to many authoritative decrees not collected in the Recopilación de Indias,[48] gave the Indian pueblos what amounted to an inherent right to available water.[49] The precise application of the royal legislation was unimportant. Indian pueblos in the Bajío were not required to show pre-conquest possession or evidence of mercedes granted at the time they were founded. Their rights were simply assumed to exist. The colonial administrators understood the royal policy that Indian pueblos should be given an allotment of available water supplies, and the policy was carried out. In the adjudication of water rights in the Cotzala Valley, for example, the Spanish haciendas were required to produce evidence of water grants, but the Indian communities, without any proof of title, were each given one surco of water.[50]

---

[42]RLI, 6:3:9.

[43]RLI, 6:3:8.

[44]RLI, 4:12:17.

[45]RLI, 4:12:19.

[46]RLI, 4:12:18.

[47]Beleña, I, p. 208.

[48]E.g., Konetzke, I, pp. 25-26. Many similar decrees can be found in the *Ramo de Reales Cédulas* of the Archivo General de la Nación.

[49]At least in the Bajío, Spanish towns or cities acquired water rights only by grant or purchase. But see Gregorio López's gloss on SP, 3:28:6.

[50]Prem, p. 61.

The extent of the water rights, held communally by Indian pueblos, was defined by the needs of the community. This accepted principle, derived from the protective legislation, was asserted in an application of Acámbaro for a water grant: "...los poblaciones, lugares y vecindades deben ser proveido de suficientes aguas para mantenerse la población."[51] The nearby Indian pueblo of Etacuaro defended successfully its right to a large enough share of water from a spring to irrigate its lands. The pueblo argued,

> Por los leyes de nuestro Reyno se ordena que las aguas sean comunes a todos los vecinos y que los virreyes y audiencias cuiden de proveer la conveniente a la perpetuidad de los pueblos.[52]

The adjudication of water rights on the basis of need was very characteristic of northern Mexico[53] and extended to Spanish as well as indigenous communities. The few cases of this sort that arose in the Bajío, all involving Indian pueblos, illustrate that the concept of need could be used to limit as well as to legitimate a pueblo's water rights. When a landowner in Querétaro challenged the water rights of the pueblo San Francisco, a commissioned judge determined that the pueblo was making little use of its irrigation system and granted the landowner a right to two-thirds of the water from the pueblo's dam.[54] In contrast, a group of Indians congregating on the hacienda of the mayorazgo in Apaseo were able to defend their diversion of water from the hacienda's irrigation system. With a rare citation of legal authority, the audiencia ordered that they be allowed to enjoy "las aguas necesarias para su cultivo y de sus huertas y arboles frutales" in conformity with Book 6, title 3, law 12 of the Recopilación de Indias.

Water rights of Indian pueblos were an important source of title for Spanish landowners despite the protective legislation.[55] Colonial legislation required elaborate formalities for the alienation of Indian land or water; it was necessary to sell the property at public auction after a judicial determination that the community would not be prejudiced by the sale and to give a public notice of sale on thirty consecutive days.[56] These formalities do not seem to have been observed in the parcelization of Indian land and water rights in Querétaro. Yet the elaborate system of water rights that evolved from this parcelization was respected in the judicial allocation of Fernández de Castro. In an Apaseo case, the audiencia ignored undisputed evidence that the formalities had not been followed in the conveyance of two days of water.[57] Whether or not the formalities were observed in San Juan del Río, the pueblo's conveyance of water rights to Pedro de Quesada was successfully relied upon as proof of legal right in later water rights

[51] AGN-M, Vol. 74, f. 80.

[52] AGN-T, Vol. 2716, Exp. 3, f. 23v.

[53] Michael C. Meyer (personal communication).

[54] AGN-M, Vol. 70, f. 14v. Also AGN-T, Vol. 534, 2a pte, Exp. 1, f. 1v.

[55] See Gibson, p. 214; Chevalier (1952), pp. 288-89.

[56] Chevalier (1952), p. 280; Beleña, I, part 2, p. 113; RLI, 4:12:16 and 6:4:18.

[57] AGN-T, Vol. 187, Exp. 2.

litigation.[58] The early compact between Hernán Bocanegra and the Indian pueblo of Apaseo is perhaps a special case: it was sanctioned by a judicial order and long possession. But it is worth noting that it lay at the basis of the system of water rights in the area.[59]

The sources of water rights that we have examined were characterized by considerable ambiguity. The audiencia often issued grants sporadically over extended periods of time with little overall coordination. The rights of Indian pueblos were of uncertain scope and effect and could evolve gradually into complex patterns. The land compositions provided little actual clarification. There was consequently a need for judicial allocations or *repartimientos* to clarify and coordinate the water rights of users sharing a common supply. Although the repartimientos generally were carried out by judges, they had much of the character of administrative proceedings. The colonial government lacked the modern separation of judicial and administrative functions. Upon petition of an interested party, the judge would summon all affected landowners, examine titles, receive testimony, and personally visit the land in question (the personal inspection was aptly called the "vista de ojos"). Thus informed, the judge would issue an order allocating the supply.

The repartimientos of water in the Bajío normally concerned the application of previously documented legal rights, but there were two exceptions. The allocation of water between the pueblo of Acámbaro and the Hacienda San Isidro may have constituted an original award of water rights to the hacienda; the case does not mention any previous agreements or conveyance.[60] A more striking example is provided by the early history of León. The documents accompanying the foundation of León in 1576 contain little evidence of irrigation,[61] but around 1600 a number of farmers began to exploit actively the network of small streams and springs around the town. They inevitably fell into competition with each other. In 1609, a landowner petitioned the alcalde mayor for a repartimiento of the Arroyo de Señora that passed through the town. The alcalde assigned four landowners irrigation turns ranging from eight to twenty-eight days and prescribed the order in which each was to receive his turn.[62] Taylor regards repartimientos of this sort as the characteristic way that water rights were acquired in other parts of colonial Mexico.

The quasi-administrative nature of the repartimientos allowed judges to allocate water on the basis of equity or the public interest. The royal legislation expressly permitted allocation of water on the basis of "justice,"[63] and the public interest was regarded an an appropriate judicial consideration.[64] These factors appear clearly in some

[58]AGN-T, Vol. 307, Exp. 2. The formalities were observed in the earlier sale of a mill in San Juan del Río. Ibid, f. 59v.

[59]Taylor regards the division of water at Apaseo as an example of a judicial allocation. Taylor (1975), pp. 201-205.

[60]AGN-T, Vol. 534, 2a pte, Exp. 1.

[61]Rodríguez Frausto, passim.

[62]AHML, caja 1606-1609, Exp. 18.

[63]E.g., RLI, 4:17:9.

[64]AGN-T, Vol. 1362, Cuad. 3, f. 91; AGN-M, Vol. 74, f. 80v; Gregorio López, gloss on SP, 3:28:9; Rodríguez de San Miguel, II, pp. 304-305.

repartimientos in the north of Mexico,[65] but they are not found in the repartimientos of the Bajío. Apart from formally documented legal rights, the judges appear to have been attentive to possession of irrigation systems and actual use of water for irrigation.[66]

It might be thought that judges exceeded their powers in awarding water rights; in theory, title to the royal patrimony could be granted only with authorization of the Crown. But repartimientos of water rights were a practical necessity that filled part of the gaps left by the uneven distribution of water grants. The repartimiento of an alcalde in the Bajío might very well be superseded by an order of another judicial officer of the audiencia,[67] but the alcalde's authority to make the repartimiento was never questioned. As Chevalier has observed, the government was "in the hands of jurists,"[68] and they were not known to denigrate their own functions. The repartimientos, however, did not confer vested rights of the kind found in English law. The sample of repartimientos we have reviewed, together with those described by Taylor and Prem, suggest that in Central Mexico repartimientos by alcaldes had an essentially provisional character, while repartimientos by judges specially commissioned by the audiencia were usually regarded with the utmost respect. But even these latter repartimientos were potentially subject to modification. The 1573 repartimiento of the Cotzala Valley was later reviewed by another commissioned judge who made certain minor changes.

Water rights, however acquired, were normally transferable, but one occasionally encounters the concept of water rights as appurtenant to land. Since water rights were widely regarded as appurtenant to land in Spain,[69] it is not surprising that the concept occurred in the New World. We have seen that water rights were seldom conveyed separate from land in sixteenth-century Celaya. As late as 1678, a lawyer argued that water rights in Apaseo were appurtenant to land.[70] The case of Querétaro suggests that urban water rights may have been generally treated as appurtenant to land. The water rights assigned by the repartimiento of Fernández de Castro related to particular plots of land,[71] and the rights to the aqueduct's water supply were similarly granted only for the use of particular houses.[72]

Agreements among neighboring landowners and testamentary dispositions formed an important part of the legal relations governing water use in the Bajío. One occasionally finds an allocation of water rights among landowners with equally dubious rights. In the absence of a third party to denounce their water usage, these agreements effectively defined the parties' rights. The owner of the Hacienda Espejo neglected to secure formal water rights to the Arroyo Mandujano before building the dam described in the last chapter; as we have seen, he left two-thirds of the water to one child and one-third to another. When the children's heirs fell into litigation, the audiencia confirmed the

[65]Michael C. Meyer (personal communication).

[66]E.g., AGN-T, Vol. 674, Cuad. 2, ff. 1191-93.

[67]In northern Mexico, few judicial repartimientos were ever reviewed by higher authority (Michael C. Meyer, personal communication).

[68]Chevalier (1952), p. 269. "Apres tout, c'étaient des juristes qui gouvernaient le pays."

[69]Glick (1972), pp. 5-26; Altamira, passim; Ruiz-Funes García, passim.

[70]AGN-T, Vol. 187, Exp. 2, f. 204.

[71]AGN-T, Vol. 455, Exp. 4, ff. 1v, 2v.

[72]MNA.AM-Qro, roll 27, Mercedes de Agua.

division established by this testamentary disposition without questioning the parties' water rights.[73] A further example can be found in León. In 1629, there was one major dam, shared by three haciendas, on the river bordering the town. Each of the three landowners had claims to water rights based on the repartimiento of 1609, mercedes or the "registration" of water rights, but the rights conferred by these documents had no clear application to the pattern of irrigation established by the dam. Accordingly, the landowners entered into an agreement providing for the division of water at the dam between two haciendas and the periodic release of water to the third hacienda located downstream. In 1700, the successors in interest to the downstream hacienda successfully vindicated their claim to a share of the water by relying on the 1629 agreement.[74]

Most water litigation in the Bajío concerned the interpretation of formal legal documents. The issues included, in addition to matters discussed elsewhere, the interpretation of grants of water rights,[75] the power of a religious confraternity to rent water rights,[76] the conclusive effect of a judicial decree,[77] the forgery of a conveyance of Indian water rights,[78] the violation of a rental agreement,[79] and the application of repartimiento orders.[80] In cases of this kind, arguments based on existing rights are nearly always interspersed with vigorous assertions of the claimant's long use of the disputed water. At times, these arguments invoke concepts with a definite legal significance, but just as often, they are vague appeals based on the landowner's long possession. The owner of a hacienda in Acámbaro typically asserted, "Siempre las [aguas] poseyeron quieta y pacificamente sin contradiccion alguna."[81] Other common formulas include "uso antiguo" and "continuada y pacífica posesión." One may be inclined to dismiss these as rhetoric or as makeweight arguments of no real consequence. But there can be no doubt that a respect for possession, derived from the study of Spanish and Roman law, was deeply engrained in the mentality of the colonial lawyer. Citing Roman law, Solórzano typically wrote:

[73] AGN-T, Vol. 1353, Exp. 1, and 1354, Exp. 1.

[74] AGN-T, Vol. 192, Exp. 1. Also AGN-T, Vol. 675, Exp. 1 (agreement); Vol. 2951, Exp. 31 (testamentary division).

[75] AGN-T, Vol. 383, Exp. 3; Vol. 544, Exp. 3; Vol. 674, Cuad. 1 and 2; Vol. 1168, Exp. 3; Vol. 1169, Exp. 1; Vol. 1175, Exp. 4.

[76] AGN-T, Vol. 450, Exp. 5.

[77] AGN-T, Vol. 2716, Exp. 3, e.g., cuad. 2, f. 2v.

[78] AGN-T, Vol. 187, Exp. 2.

[79] AGN-T, Vol. 2901, Exp. 36.

[80] AGN-T, Vol. 764, Exp. 3; Vol. 2959, Exp. 141; Vol. 2967, Exp. 112; MNA.AH-Qro, roll 34, Arch. Not. 4, leg. civ 1705-1709 and leg. civ 1668-1694 (barrios San Sebastián, San Roque, and San Francisco).

[81] AGN-T, Vol. 1446, Exp. 6, f. 8.

> La posesión antigua ha de vencer, prevalecer y ser amparada porque la nueva o posterior se presume violenta y clandestina.[82]

In many cases with an independent basis for decision, the previous patterns of use may well have influenced the outcome of the case.[83]

Apart from such vague appeals, four doctrines of possessory rights played a role in water cases: immemorial possession, prescription, possession for a year and a day, and usurpation of possession or *despojo*. No legal term of art appears so frequently in the water cases of the Bajío as "immemorial possession." The concept had roots in Roman law[84] and diverse applications in Spanish law.[85] It was said to remedy defects in title with the presumption that legitimate title had been lost.[86] But it also had a very precise application to water cases. Partida 3, title 31, law 15 provided:

> Mas las otras servidumbres de que se ayudan los homes para aprovechar y labrar sus heredades et sus edeficios, que no usan dellos cada dia, mas a las veces et con fecho, asi como...ó en agua que veniese una vez en la semana o en el mes o el ano et non cada dia, tales servidumbres como estas y las otras semejantes dellas...ha meester que hayan usado dellas ellos ó aquellas de quien las habieran tanto tiempo que non se puedan acordar los homes quanto ha que lo comenzaron a usar.[87]

Most water cases involved just such discontinuous uses. In addition to numerous invocations of the concept in arguments, one finds serious efforts to prove it. In the Valle de Santiago, witnesses were asked if a hacienda had irrigated from a secondary watercourse as long as they could remember.[88] The concept unquestionably entered into the judicial decision of certain water cases.[89]

Accepted legal doctrines contained two schools of thought in construing the concept of immemorial possession: one applied it literally; the other took it to mean one hundred years. In his widely read commentary on the Siete Partidas, Gregorio López discussed these two interpretations and indicated that the concept of immemorial

---

[82]Solórzano Pereira, I, Book 2, Chap. 23, No. 6.

[83]AGN-T, Vol. 307, Exp. 2; Vol. 1175, Exp. 4.

[84]Dobkins, pp. 52-53; Lasso de la Vega, pp. 4-5 (citing commentaries on Roman law).

[85]See gloss of Gregorio López, SP, 3:29:7 and 3:31:15; Pérez y López, XXIV, pp. 4, 26.

[86]AGN-T, Vol. 764, Exp. 3, Cuad. 1, f. 111v.

[87]See also Galván Rivera, p. 15.

[88]AGN-T, Vol. 586, Exp. 8, f. 37.

[89]AGN-T, Vol. 383, Exp. 4; Vol. 988, Exp. 1, Cuad. 3, f. 83; Vol. 1169, Exp. 1, Cuad. 4, f. 11v, Cuad. 5, f. 12.

possession as one hundred years was upheld by good authority.[90] This interpretation was strictly applied by the chief legal advisor of the audiencia in a Querétaro case decided in 1753. He found that the claimant's rights were based on the repartimiento of 1654 and that his use was interrupted in the year 1723. Consequently, he reasoned:

> ...asi por esta interrupcion como por el reciente origen de la posesion de dicha Bachiller y sus causantes resulta que dicha posesion es menor que centenaria y que no puede ser immemorial.[91]

The audiencia followed his recommendation to reject the claim.

The laws of prescriptions in the Siete Partidas established the general rule that a person could acquire title to property with ten years possession, if the previous owner resided on the land, or with twenty years possession, if the owner were absent.[92] Partida 3, title 31, law 15 specifically applied this principle to a continuous flow of water.

> De tal natura seyendo la servidumbre que ficiere á otri cotidianamiente sin obra de aquel que la rescibe, asi como si fuese aguaducha que corriese de fuente que rasciese en campo de alguno o otra semejante della, si el vecino sirviese desta agua regando su heredat diez años estando su dueño en la tierra et non lo contradiciendo o viente seyendo fuera de ella et esta ficiese á buena fe.

An applicant for a water grant in León placed himself squarely within the provision of this law. He alleged that, while he lacked title, he had used a spring to irrigate his land for ten years and petitioned that his prescriptive water rights be confirmed by a formal grant.[93] But the right to acquire title by prescription was hedged with limitations. The use had to be continuous, in good faith, under color of title, and for the prescribed period.[94] In part because of these limitations, the claim of prescription was not often raised in water cases in the Bajío.[95] The issue was seriously litigated in only one case and the claim was rejected because of an insufficient period of time.[96]

The closely related doctrines of possession for a year and a day and usurpation of possession (despojo) were of particular importance in eighteenth century litigation among haciendas lacking formal water rights. The concept of possession for a year and a day

---

[90]See Gregorio López, gloss on SP, 3:31:15; NR, 4:15:1 and 10:17:1. But in northern Mexico the phrase was sometimes applied simply to prior use, even where the prior use was for a brief period (Michael C. Meyer, personal communication).

[91]AGN-T, Vol. 764, Exp. 3, Cuad. 1, f. 113v.

[92]SP, 3:29:8 and 3:32:14.

[93]AGN-T, Vol. 2701, Exp. 26, f. 1.

[94]Galván Rivera, p. 11; AGN-T, Vol. 353, Exp. 2, f. 15; Vol. 764, Exp. 3, f. 114.

[95]But see AGN-T, Vol. 192, Exp. 1, f. 110v; Vol. 675, Exp. 1, f. 81; Vol. 1107, Exp. 1, f. 151; Vol. 1354, Exp. 1, f. 35.

[96]AGN-T, Vol. 2767, Exp. 3.

originated in the *Fuero Viejo de Castilla* where it had various applications.[97] It was incorporated in a rather cryptic provision of the Nueva Recopilación, later affirmed by the Novísima Recopilación.

> El que poseyere casa, viña o heredad, por año y dia en paz y faz del que se la demanda, no tiene obligación a responder sobre la posesión, teniendo titulo y buena fe; lo que se entienda de los pueblos donde haya tribunal para prescribir por año y dia.[98]

Galván Rivera stated the principle in 1842.

> La posesión se prescribe por año y un día; esto es, que el que tiene una cosa por dicho tiempo, con titulo y buena fe en paz y en faz de quien la demanda puede escusarse de responder sobre su posesión.[99]

Possession thus became legally effective for some purposes after a year and a day. The importance of this principle in water cases lay in the protection accorded to possession during the pendency of lawsuits.

The Nueva Recopilación provided that one could not be deprived of possession without being first given an opportunity for a legal hearing: "Ningun Juez despoje, ni otro, a persona sin oirla y vencerla."[100] It was a well-established procedural rule that in cases involving issues of possession (*tenencia*) and of title (*señorio*) the question of possession was to be decided first.[101] The protection accorded to a person in possession was fortified by the related principle that a person forfeited his rights by forcibly dispossessing another. The Nueva Recopilación provided:

> El que entrare o tomare alguna cosa que otro tenga en su poder y en paz, si tuviere derecho pierdalo, y si no lo tiene restituyala con otro tanto o con la valia.[102]

These rules were more than procedural provisions; they actually provided a legal cause of action. Solórzano explained,

> ...no le incumbe probar, ni exhibe titulos sino solo haver poseido y estar despojado...en probando esto ha de ser restituido ante todas cosas.

---

[97] Pérez y López, XXIV, p. 11 et seq.

[98] NR, 4:15:3. Also NovR, 11:8:3; Pérez y López, XXIII, p. 257, XXIV, p. 27.

[99] Galván Rivera, p. 13.

[100] NR, 4:13:2. Also AGN-T, Vol. 1169, Exp. 1, Cuad. 5, f. 68; Vol. 2072, Cuad. 8, f. 62v.

[101] SP, 3:2:27.

[102] NR, 4:13:1. Also, AGN-T, Vol. 675, Exp. 1, f. 81v; Vol. 2767, Exp. 3, Cuad. 5, f. 84.

And again:

> La posesión, y mas quando es continuada por algunos años, es tan poderosa, que debe ser uno amparado y manutenido en ella por solo titulo aparente, aunque no sea concluyente, hasta que ventilado eso con mas espacio, sea vencido en el juicio de la propiedad.[103]

A person needed only to allege and prove that he was forcibly dispossessed to secure an interlocutory order restoring him to possession. Proof of possession for a year and a day was sufficient to establish a usurpation of possession (despojo) that entitled the landowner to this provisional remedy. An unauthorized use thus would qualify for protection after that period. As stated in a Querétaro case: "...y en caso de que aya pasado año y dia desde el dia del despojo no innove y remita las diligencias..."[104] To prevent this from happening, the haciendas sharing the canal of Gogorrón in Salvatierra closed off all unauthorized outlets once a year.[105]

In eighteenth century lawsuits between haciendas lacking legally documented water rights, the issue of usurpation of possession (despojo) frequently presented the legal battleground on which water rights were actually litigated.[106] The provisional remedy could have decisive importance where neither party had good title to water. This was well illustrated by a Celaya case. The Hacienda del Molino, as we have seen, drew water from a swamp on the far side of the city. When a landowner bordering the swamp blocked the outlet and diverted water to other property, the owner of the Hacienda del Molino, Crespo Gil, claimed a despojo. Both parties could rely only on an obscure and inconsistent history of use to sustain their rights. The case came before Miguel Domínguez who stated that he was called upon to decide only two issues.

> Si Crespo estaba en posesion que refiere por mas de año y dia y si con efecto fue despojado sin ser por fuero y derecho vencido.[107]

He found both questions in favor of the Hacienda del Molino and expressly reserved other issues for a determination by the audiencia. With precisely the same legal strategy, Crespo Gil defended his possession of a dam near his hacienda and headgates on the River Laja that were destroyed in 1806 by order of the ayuntamiento. Both cases continued for years; the documentary record ends with the advent of independence, but in the twentieth century the Hacienda del Molino still retained an irrigation system.[108] We may surmise that the provisional orders actually provided a decisive victory.

---

[103]Solórzano Pereira, II, Book 3, Chap. 3, para. 27.

[104]AGN-T, Vol. 764, Exp. 3, f. 72.

[105]AGN-T, Vol. 2747, Exp. 1, ff. 37v, 52.

[106]E.g., AGN-T, Vol. 533, Exp. 2, f. 48; Vol. 675, Exp. 1; Vol. 1107, Exp. 1, f. 86v; Vol. 1113, Exp. 1, Cuad. 5, f. 45; Vol. 1168, Exp. 3, f. 60v; Vol. 1353, Exp. 1, f. 94v; Vol. 2767, Exp. 3; Vol. 3618, Exp. 1.

[107]AGN-T, Vol. 1362, Exp. 1, Cuad. 3, f. 86v.

[108]AGN-T, Vol. 1390, Exp. 3; Vol. 2071-73; Vol. 2074, Exp. 2, f. 36; Velasco y Mendoza, I, p. 18.

The widespread use of possessory concepts in water cases could coexist with the theory of water rights as dispensations of the Crown because the concepts aided in the practical resolution of cases. The theoretical underpinnings of the two systems were indeed inconsistent: the concept of water as part of the royal patrimony implied that it could be alienated only by the authority of the Crown,[109] whereas possessory rules presupposed alienability by private action. The inconsistency existed despite the inheritance of Roman law which distinguished between public and private waterways. Generally, under Roman law, public rivers were perennial and private waterways were intermittent.[110] This distinction was known in scholarly circles in colonial Mexico and was explicitly embraced by Lasso de la Vega.[111] Even Solórzano in his sweeping affirmation of the Crown's rights speaks of "ríos públicos," a term taken from Roman law.[112] But I have found scarcely any trace of the distinction in the cases.[113] The use of possessory legal theories apparently did not represent a recognition of the private character of waterways. Neither did it represent a weakening of the concept of water belonging to the Crown; the concept of the royal patrimony was freely invoked when it helped to justify the exercise of judicial authority or aided in the resolution of a controversy. The widespread use of possessory legal theories evidently reflected no more than the practical value of these considerations in resolving questions that could not easily be decided under the system of administered rights.

An effective private control over water could be asserted by other means. In his treatise on water measurement, Sáenz de Escobar observed: "Por ser regla general aunque qualquier puede obrar libremente en sus tierras no debe ser en perjuicio de tercero."[114] Since Sáenz served both as legal advisor and oidor of the audiencia, his comment reflected the actual administration of the law. The theory represented the converse side of the principle of "sin perjuicio de tercero." While the principle itself limited private water rights, it had the corollary that private landowners could make any use of water on their land so long as others were not adversely affected. The corollary served to justify private appropriations. This principle that one could make free use of his own land so long as it did not prejudice others seems to have rested on a kind of jurisprudential generalization rather than on specific legislation. It may well have had considerable importance in other regions characterized by numerous small private appropriations of water, but it appears in only a few Bajío cases. In Celaya, the alcalde ordinario allowed the owner of the Hacienda del Molino in 1795 to present evidence that his use of the headgates and canal on the River Laja did not prejudice other landowners.[115] The heirs to the Hacienda Espejo in Querétaro also argued the theory at great length, but it does not appear to have governed the actual decision in the case.[116]

---

[109]SP, 3:29:6, 7 and gloss of Gregorio López on the law; Sáenz, *Libro de las Ordenanzas* (mss), f. 73v; Lasso de la Vega, p. 2.

[110]Spota, pp. 119-88.

[111]Lasso de la Vega, p. 4. See also pp. 2, 6.

[112]Solórzano Pereira, V, Book 6, Chap. 12, p. 37.

[113]But see AGN-T, Vol. 3618, Exp. 1, Cuad. 2, f. 29.

[114]Sáenz, *Libro de las Ordenanzas* (mss), f. 65.

[115]AGN-T, Vol. 2767, Exp. 3, Cuad. 5, ff. 84, 92-93v.

[116]See AGN-T, Vol. 1353, Exp. 1, e.g., f. 119; Vol. 1354, Exp. 1, e.g., f. 163v.

The strongest case for private control of water could be made in the case of springs. The provisions of the Siete Partidas relating to springs presuppose that a landowner had a right to the water of springs originating on his land. For example, Partida 3, title 31, law 5 provided:

> Ganada habiendo home servidumbre de traer agua para regar su heredamiento de fuente que nasciese en heredat agena si despues el dueño de la fuente quisiese otorgar á otri poder de aprovecharse de aquella agua, non lo puede face sin consentimiento de aquel a quien primero fue otorgada la servidumbre della, fueros ende si el agua fuese tanta que abondase al heredamiento de amos.[117]

Drawing on this tradition, Lasso de la Vega held that grants of land carried with them the right to springs located on the land.

> Pero si en la concessión de las tierras se concede juntamente las aguas sus originales, por considerarse partes o fructos de dichas tierras mercenadas es doctrina del padre Avendaño nuestra Regnicolo en su Thesauro Indica.[118]

But this doctrine conflicted with a royal decree containing the most uncompromising assertion of the Crown's rights to the waters (las aguas) of the Indies.[119] The few documents on point reflect this inconsistency in the law. A judge reviewing the water rights of a hacienda in San Miguel El Grande clearly indicated that the title to the land conferred a right to springs originating thereon.[120] But other documents, including the recitation in a grant of water rights in Celaya, reflected the opposing point of view.

> Siendo como son pertenecientes al Real Patrimonio y competandose entre las cosas que son de regalia los manantiales de agua aunque emanen y tengan su nacimiento en tierras mercenadas o otras terceras se sigue que la donacion hecha a esta parte por el dueño del fundo es inutil por careser el donante de dominio en la cosa donaba.[121]

The two interpretations came into conflict in a San Juan del Río case.[122] A landowner claimed the right to dam an arroyo since it originated on his property, but the audiencia required him to release a part of the water to a downstream hacienda.

---

[117]Also SP, 3:31:12 and 3:32:19.

[118]Lasso de la Vega, p. 5.

[119]E.g., RLI, 4:17:5.

[120]AGN-T, Vol. 1169, Exp. 1, Cuad. 5, f. 108. Also AGN-T, Vol. 192, Exp. 1, f. 101.

[121]AGN-M, Vol. 72, f. 269. To same effect: AGN-M, Vol. 81, f. 130; AGN-T, Vol. 2765, Exp. 11, f. 11.

[122]AGN-T, Vol. 272, Exp. 1.

Under the Siete Partidas, the diversion of water from its natural course could result in liability. For example, Partida 3, title 32, law 13 provided:

> Si alguno alzase pared ó ficiese estacada o valladar ó otra labor en su heredat, de guisa que el agua non podiese correr por el logar por do salie, por que se habiese hi de facer estanque de que veniese daño a los heredades que son vecinas; o si por aventura alzase alguna labor en logar por do solie el agua venir et por aquel alzamiento se mudase el curso della et cayese de tan alto que ficiese foyas ó cavas en heredat de su vecino, o la embargase o detoviese el agua de guisa que los otros que la solien haber non podiesen regar sus heredades della asi como solien...debe seer derribada á su costa et á su mision et tornada al primo estado.

This principle, elaborated in several other laws of the Partidas,[123] was invoked in several water cases, sometimes with precise citations, to challenge another landowner's construction of a dam.[124] The clearest application of the principle appears in a dispute between the pueblo San Francisco del Rincón and a neighboring hacienda.[125] The pueblo had long maintained an earthen dam to irrigate its fields, but, by improving the dam, the Indians flooded a neighboring hacienda. The alcalde mayor recognized their obligation to remove the dam, but sought to reduce the burden of doing so. He ordered that the next time the dam was destroyed by high water it should be rebuilt so that it would not flood the hacienda's land.

Colonial New Spain inherited the Roman law of easements, as incorporated in the Siete Partidas. Water easements were of two kinds: natural and conventional.[126] The former was created by the natural flow of water over a landowner's land onto neighboring land;[127] the latter could be created by contract, testamentary disposition, or prescription.[128] In theory, the existence of a valid easement turned on technical considerations, but attempts by landowners to block the passage of water to neighboring land do not not seem to have been much favored, at least if we may judge from the fragmentary record of two Bajío cases concerning such attempts.[129] The Siete Partidas, as construed by Gregorio López, also stated two general principles with respect to the maintenance of ditches: (1) a person holding a right to bring water across another's land was responsible for maintaining the ditch or conduit;[130] (2) ditches shared by

---

[123]SP, 3:32:15, 16, 17, and 18. Also NovR, 7:26:7.

[124]AGN-T, Vol. 2073, Exp. 1, f. 33v; Vol. 2963, Exp. 49; Vol. 3618, Exp. 1, Cuad. 2, ff. 23, 30. Also Lasso de la Vega, p. 9.

[125]AGN-T, Vol. 1124, Exp. 1.

[126]AGN-T, Vol. 1354, Exp. 1, f. 34; Sáenz, *Libro de las Ordenanzas* (mss), f. 65.

[127]SP, 3:32:14.

[128]SP, 3:31:14; Lasso de la Vega, p. 9.

[129]AGN-T, Vol. 353, Exp. 2; Vol. 586, Exp. 1.

[130]SP, 3:31:4.

neighboring farms were to be jointly maintained.[131] Local practices, despite certain variations, were always consistent with these very general principles.

In his interesting history of the irrigation system at San Antonio, Texas, Thomas Glick argues for the importance of local custom rather than the formal law of water rights in the actual practice of irrigation.[132] An example of such a custom is found in León. In the early seventeenth century, landowners engaging in irrigation invariably followed the practice of "registering" water rights with the municipal government.[133] It is unclear what effect this registration had under the law of the Indies. At least one landowner, evidently doubting its efficacy, had his registered water right confirmed by the audiencia.[134] The practice later died out, but it is of interest because it suggests that more forms of local regulation of water rights may have existed than appears in surviving documentation.

Of greater importance was the distinction between primary and secondary water rights, which was very well established in actual practice but did not appear in formal law.[135] Landowners often held a right to the remanientes (sometimes known as the *sobras* or *demasias*) or the water remaining after an upstream landowner had irrigated his land. Given the technology of the time, the concept of the remaniente could not be avoided; there was frequently no other way to describe the shares of water enjoyed by two farms or by a pueblo and a neighboring farm. Water grants and repartimientos intended to legitimize existing uses necessarily resorted to the term.[136] But the concept of the remaniente was vague and led to disputes. In one case, it was argued that the concession of a right to the remaniente of water limited the person holding the prior water right to no more than normal irrigation of his property.[137] In the absence of any formal law on the question, the argument was inconclusive.

With the cession of northern Mexican territories to the United States in the Treaty of Guadalupe Hidalgo, the colonial water law of Mexico remained applicable to existing water rights. The litigation of these rights has had great social importance in the southwestern United States, but it has not led to an accurate understanding of colonial water law. As a system of adminstered rights, colonial water law embodied concepts having no parallel in the Anglo-American system of law. Moreover, the keystones of this system--the royal assertion of dominion over water and the protective legislation regarding Indian pueblos--were only aspects of the actual operation of the law. The practical adjudication of water rights required the use of judicial repartimientos and the importation of principles of Spanish law, both of which introduced new legal considerations. This unexpected complexity of colonial water law has never been appreciated by the U.S. courts.

---

[131]Gloss of Gregorio López, SP, 3:31:4 and 3:32:16.

[132]Glick (1972), p. 3.

[133]AGN-T, Vol. 192, Exp. 1, ff. 55v, 72v, 101-36; AHML, caja 1613-14, Exp. 13, caja 1617-19, Exp. 10; caja 1646-48, Exp. 20; caja 1660-63, Exp. 4.

[124]AGN-M, Vol. 31, f. 172.

[135]E.g., AHML, caja 1685-90, Exp. 3; caja 1613-14, Exp. 13; AGN-T, Vol. 580, Exp. 1, Cuad. 1, f. 298; Vol. 1107, Exp. 1, f. 147v; Vol. 1446, Exp. 6, f. 6; Vol. 2970, Exp. 107.

[136]AGN-M, Vol. 31, f. 238; Vol. 32, 101v; Vol. 34, f. 14v; AGN-T, Vol. 2648 (numerous examples).

[137]AGN-T, Vol. 764, Exp. 3, Cuad. 3, f. 64. Also AGN-T, Vol. 192, Exp. 1, f. 130.

One of the most curious products of the U.S. litigation of Mexican water rights has been the notion that colonial law contained the doctrine of riparian rights. The notion stems from the *Diccionario Razonado de Legislación* of Joaquín Escriche, a Spanish legal writer of the nineteenth century. After an exposition of the laws of the Siete Partidas relating to water, most editions of the text proceed to expound the doctrine of riparian rights. Two early editions published in Mexico and edited by Rodríguez de San Miguel, the most distinguished Mexican lawyer of his day, are, however, among those that lack this statement of riparian rights.[138] The relevant section of Escriche's text thus was probably unknown, or virtually unknown, in Mexico before the cession of the northern territories, but it was unfortunately translated into English in a nineteenth century legal treatise and passed on to the U.S. courts. It can be readily dismissed as an attempt of the author to interpret Spanish law in conformity with the Napoleonic Code.[139] One may, of course, argue that the practical consequences of a theory of riparian rights are not much different from those of colonial water law.[140] Riparian landowners lacking grants of water might secure rights in a repartimiento or defend an unauthorized use on theories of possession; Indian pueblos could similarly claim a share of neighboring water courses under the protective legislation. But the concept of water rights implied from the grant or ownership of riparian land never appears in colonial law and would have been most alien to the mentality of the colonial lawyer.

A similar difficulty underlies Taylor's contention that some water disputes were decided on the basis of "proximity to source."[141] He cites two cases in which upland property owners were able to control the use of streams originating on their land. Upstream users are always in an advantageous position to usurp a disproportionate share of water, and under colonial water law, they had some opportunity to defend their appropriation of water through theories of possession, sin perjuicio de tercero or the ownership of springs. But "proximity to source" as a legal doctrine in adjudicating water rights was unknown to colonial law.

Colonial water law reflected the centralized, legalistic but loosely administered character of colonial government. In modern times, the importance of administered rights has been greatly enhanced. Virtually all the water resources of Mexico are now declared to be "property of the nation." Water rights are obtained by applying to the Secretaría de Recusos Hidráulicos, which grants concessions for limited periods and which are subject to conditions.[142] The administration of water rights thus has been tightened but the system itself has a basic continuity with that of the colonial period.

Conclusion

While in some societies close linkages between the practice of irrigation and patterns of social organization are revealed by anthropological and historical studies,[1] such linkages were of decidedly minor importance in the colonial Bajío. Irrigation does not appear to have created onerous demands for labor; the early repartimientos of Celaya mention field work, not irrigation; other documentary sources allude only to the annual deployment of peons to clean ditches; the account of the construction of the dam on the Hacienda Mendoza reveals that it was built in a single dry season with labor that was abundant at that time of year. Neither did the conduct of irrigation require much in the form of a "responsbile authority."[2] In Salvatierra, no such authority existed; in Celaya, it existed during the sixteenth century but largely disappeared thereafter; in Querétaro, the hydraulic bureaucracy consisted of a maintenance employee (fontanero) and a few town officials whose duties occasionally involved the regulation of water use. Only in the Valle de Santiago was there a water judge with relatively important functions in the supervision of irrigation practices. It seems idle to ask other questions posed by the anthropological literature: was there a connection between water control and social power? Was the administration of irrigation tied to social stratification? Did irrigation strengthen the role of the central government? Nowhere--not even in the Valle de Santiago--did the practice of irrigation vitally affect social or political structures. This finding comes as no surprise in the case of the irrigation systems of Celaya, Salvatierra, and the Valle de Santiago. They were, after all, composed of small groups of essentially commercial enterprises, the wheat-growing haciendas. One would not expect the same nexus between irrigation and social organization as in large-scale irrigation systems or communities of small irrigators. But the case of Querétaro is of genuine interest for the student of cultural evolution.

Querétaro occupied a hydraulic niche commonly found among pre-industrial cities in semi-arid regions. Its prosperity depended on the husbanding and elaborate distribution of a small supply of water to an urban community and a surrounding area of gardens and farms. We have seen that the elaborate system of water distribution in Querétaro evolved incrementally in the seventeenth century with no central planning or adminstration and, after being confirmed in the repartimiento of 1654, was maintained by custom and occasional resort to judicial remedies. The expansion of the system in the eighteenth century did require the intervention of a kind of higher power, but it took the form of the private benefaction of the Marqués de Villar de Aguila rather than state

[1]E.g., Steward, passim; Wittfogel, passim; Gray, passim; Beardsley, passim; Downing and Gibson, passim; Johnson, passim.

[2]The phrase is that of John W. Bennett in his comment on an article of Robert C. and Eva Hunt. Hunt and Hunt, p. 398.

action. Thereafter, the management of the water system was no more than one of many concerns of municipal officials. If the water system of Querétaro demanded only a minimal degree of governmental intervention, we may ask why other pre-industrial cities with similar hydraulic needs would require more. The question is pertinent because there have been serious attempts to apply the "hydraulic hypothesis" for the creation of the state to such cities as Teotihuacan and Cuzco in pre-Hispanic America.[3] These cities were larger than Querétaro and far removed from it in cultural development, but they faced similar problems of water management with pre-industrial technologies. Unless one can completely distinguish the cases, the history of Querétaro provides reason to look for factors other than hydraulic management as explanations for the evolution of political power in these urban centers and others with similar ecological setting.

While the practice of irrigation had only minor impact on social organization, the social and political structures of the colonial period conditioned the exploitation of water resources in many ways.[4] It was the individual initiatives of the landowning elite and sporadic interventions of central authority that most consistently affected the history of irrigation. Municipal government, subordinate to the central government and hobbled by inadequate finances and legal disabilities, could not easily become a focus of cooperative endeavors. Even in the Valle de Santiago, where the municipal role was most important, written contracts among landowners, rather than municipal authority, came to be the key instrument of cooperation in the eighteenth century. The peculiar forms of hydraulic exploitation in the colonial Bajío, as in other regions, were largely the product of the interplay between the resources and authority of a landowning elite and the loosely administered, but authoritarian, rule of the central government. The point has been carried further by certain writers of an evident ideological persuasion.[5] Eva and Robert Hunt see signs of a specifically colonial form of water management tending to reinforce the power of the dominant class.

> Procedures to convert illegal water allocations into new legal codes for water distribution favoring haciendas and mills were highly standardized. They involved, among other things, the recurrent bribing of witnesses, the stealing of wooden water measuring templates, the control of lawyers in the capital, local military harassment, and threats of economic sanctions against villages...One point which needs investigation is the impact of colonial systems on the management of irrigation systems.[6]

The Bajío with its widely scattered Indian pueblos is not the best area to test this opinion, but our study of water law suggests a more qualified view. The administration of water rights was always mediated by a complex and benign system of law that not infrequently served to check private usurpations of water. Despite probable political manipulations, the sporadic interventions of central authority often tended to temper, rather than to reinforce, the social dominance of the Spanish landed class.

In considering the control of water resources, we face the preliminary question of whether water rights evolved independently of the control of land. Chevalier suggests

---

[3]Sanders and Price, p. 104-105; Price, p. 55; Villanueva U. and Sherbondy, p. XVIII.

[4]Compare: Geertz, passim.

[5]E.g., Frank, p. 67.

[6]Hunt and Hunt, pp. 395-96.

that there may have been a struggle over water that paralleled the struggle for land, but he reserves the question for further research. The answer, he says, will require lengthy investigation ("longue recherches").[7] We have seen that, while water rights were indeed an independent variable in the evolution of land tenure, the control of water usually followed closely on the control of land. Owners of irrigable land, lacking water rights, were only occasionally deterred from developing irrigation systems. The water law of the colonial period acquired a complexity that offered the landowner many avenues to secure formal water rights or to defend an unauthorized use of water that did not encroach on the superior rights of a third party. Moreover, water rights were commonly transferred in connection with the conveyance of land. Only in Celaya was there something resembling a separate market in water rights, and, even there, one finds only a few examples of individual efforts to accumulate them. Colonial litigation over water rights usually called for no more than adjustments in the patterns of water use among neighboring landowners. A single case in seventeenth century León involved a dramatic, but largely unsuccessful, attempt to monopolize water rights on a large scale.[8]

But did the practice of irrigation itself affect the evolution of land tenure? Irrigation has been linked with the emergence of latifundia in Peru, but Borde and Góngora found that irrigated properties in the Puangue Valley in Chile actually tended to be of medium size; the largest estates and the smallest parcels were most often found on unirrigated land.[9] Within the Bajío, the inadequate early allotment of water in Celaya stimulated the consolidation of properties, but elsewhere, especially in Querétaro and Apaseo, one finds an abundance of small irrigated properties. In the Valle de Santiago, some of the first small irrigated haciendas persisted throughout the colonial era, and, even in Celaya, the pattern of landholding that eventually emerged was one of medium-sized haciendas. In general, irrigation seems to have favored somewhat the survival of small properties. We may surmise that the higher productivity of irrigated land discouraged its monopolization in a few hands by reducing the incentives and increasing the cost of land acquisition. But the distinctive features of land tenure in these irrigated areas were only minor variations on a theme; the similarity to the prevailing pattern is most striking. The irrigation communities of the early Bajío were molded by the same forces that shaped the evolution of land tenure elsewhere. The fact that these communities, despite their unique origin and form of agriculture, should in the end become aggregations of haciendas similar to those in other agricultural areas reveals how pervasive were the cultural and economic influences that gave rise to the hacienda system.

Throughout this essay we have sought to establish the chronology of the construction of waterworks in the Bajío. The most significant finding has been that the major irrigation systems were established in the sixteenth century or the first quarter of the seventeenth century. It has been more difficult to trace investment in waterworks later in the colonial period; the construction of dams and storage basins was usually not accompanied by formal grants of water rights and is not always revealed by surviving estate inventories and other documents; the region of Silao and Irapuato, though apparently a favored area for agricultural investment by Guanajuato miners in the eighteenth century, offers the most meagre documentary record and has remained largely beyond the reach of our study. But enough investment in waterworks can be documented

---

[7]Chevalier (1952), pp. 184, 288. Chevalier's references to individual accumulations of water rights in the Bajío had little significance; both involved attempts to salvage some value from failing irrigation systems that soon afterwards disappeared. See AGN-M, Vol. 14, f. 178; Vol. 35, f. 73v.

[8]See AGN-T, Vol. 192, Exp. 1, e.g., 110v.

[9]Borde and Góngora, pp. 105, 118-25; Mariátegui, p. 75.

or inferred from tithe data that one can perceive a chronological pattern: the establishment of irrigation systems below convenient diversion points on the Rivers Lerma and Laja or near plentiful springs occurred in the earliest period of settlement; the exploitation of other less accessible sources of irrigation through the construction of dams and storage basins, both on the major rivers and small arroyos, proceeded actively in the eighteenth century.

But what can be said of investment in waterworks in the last three quarters of the seventeenth century? In a beautifully written and insightful essay, Andrés Lira and Luis Muro take issue with Borah's description of the seventeenth century as a "century of depression." They acknowledge the general absence of monumental constructions in the century, but they insist

> ...there is something else that has not been charted: the development of less monumental improvements; the transformation of hamlets into cities and towns; the waterworks that permitted irrigation with water drawn from dams and artificial reservoirs, and other less monumental and conspicuous constructions, affecting everyday life, which were undertaken and maintained in the era in question and which, because of their common and ephemeral nature, are more difficult for the historian to discover and evaluate than the monumental works attributed to the sixteenth and eighteenth centuries.[10]

Our study of waterworks provides only modest support for this speculation. The increasing Spanish population of the towns did stimulate the construction of an aqueduct in Celaya and a succession of improvements in Querétaro. But elsewhere one cannot find any firm evidence of an upward trend of investment. The few improvements, such as the masonry dam at Celaya, may have been offset by the deterioration of other facilities.

The value of haciendas in the Bajío depended on the quality of the land, the proximity to markets, and the investment in land clearance, fences, barns, residential buildings, and waterworks.[11] A certain minimal investment in waterworks was necessary for the land to have much value: the animals and human population needed a perennial supply of water to survive the dry season. But when that basic necessity was satisfied by the construction of norias, wells, or small dams, further hydraulic investment was merely one of several avenues a landowner could pursue to add value to his land. What importance did such investment in waterworks have in the economic history of the hacienda? The question can be best pursued by studies, such as those of Bazant and Brading,[12] that comprehend all aspects of the hacienda's economy. Our study of water systems does not permit us to analyze the importance of hydraulic investment within a broader economic context, but it has produced some useful data. We have seen that it was the desire to shield the wheat crop from fungal infestation, not the pursuit of higher productivity, that ordinarily attracted major investments in waterworks. The demands on labor and capital varied with the technology employed. Some systems exploited convenient springs or diversion points on rivers; other required substantial dams and storage basins; some needed to be at least partially rebuilt every year; others demanded only periodic maintenance and the annual cleaning of ditches. We have found no

---

[10]Lira and Muro, p. 97.

[11]Morin, pp. 242-43.

[12]Bazant, passim; Brading (1978), passim.

evidence, however, that any of the colonial technologies involved heavy labor costs, and, as noted earlier, some indications to the contrary. The information on the construction and appraisal of waterworks that we have encountered, though insufficient for an economic analysis, shows that waterworks were a moderately important field of capital investment. The hacienda of the Ojo Zarco, which had waterworks appraised at 6,700 pesos out of a total valuation of 28,000 pesos, was probably typical. Only the cost of an urban supply system, the Querétaro aqueduct, equaled the level of major investments in mining.

These reflections bring us to the end of our study. We undertook the study in the hope that it would yield insight into various aspects of cultural history. It has succeeded in its purpose if, while providing no key to historical understanding, it has offered occasional glimpses into the complex reality of the colonial Mexican Bajío.

Appendix

D. Basilio Rojas loaned me from his private library a mimeographed document dated May 7, 1921, which contains excerpts from early municipal records of Salamanca, now destroyed, and various other data regarding water rights in the Valle de Santiago drawn from private sources. The excerpts from the municipal records appear to be authentic; minor details can be corroborated by land grants in the ramo de Mercedes of the Archivo General de la Nación. The accuracy of other parts of the document is less certain. The portions of the document relating to the colonial history of the Valle de Santiago are set forth *verbatim* below.

***

Los RR. PP. Agustinos adquirieren en 1540 por donación de don Alonso de Sosa, propriedades en la hacienda de San Nicolás, a la que pertencio la fracción que hoy se llama La Zanja y provisionalmente en 1594 tuvieron derechos de aguas, que mas tarde les fueron confirmados en el ano de 1606 por el Virrey don Juan de Mendoza y Luna, Marquéz de Montesclaros.

***

Pero está fuera de toda duda que, los derechos del Valle de Santiago, prevalecían en aquellos pasados tiempos con majores signos de propiedad, como lo prueban los extractos de documentos relativos a la fundación de Salamanca, en cuyas actas de cabildo hace mas de tres siglos, hay unos pasajes que textualmente dicen..."En la villa de Salamanca, en veintesiete dias del mes de abril de mil seiscientos y siete años, estando juntos y congregados como lo tienen de uso y costumbre la Justicia, Cabildo y Regimiento, convienen a saber...que por cuanto Juan Fernández vecino de esta villa y persona a cuyo cargo esta la saca del agua del cerro de Uruétaro y Camémbaro y por otro nombre el Valle de Santiago, donde esta la saca del agua del río grande con la que se puede regar...

..."En la villa de Salamanca, en veintiocho dias del mes de mayo de mil seiscientos y siete años...para hacer repartimiento de las tierras del Valle de Santiago, del riego donde esta la acequia hecha, y el agua sacada conforme al auto$de otras para el dicho efecto..."

...En la villa de Salamanca de esta Nueva Espana en seis dias del mes de mayo de mil seiscientos y nueve años...Dijeron que mandaban y mandaron...y si hubiese quien quiera alegar las acequias del Valle de Santiago jurisdicción de esta villa donde esta sacada el agua con que riegan las labores que hoy tiene y para adelante vayan corriendo...

...En la villa de Salamanca, en primero día del mes de septiembre de mil seiscientos y nueve años...mandaron entre otras cosas que todos los vecinos que tienen labores en el Valle jurisdicción de esta villa, tengan todas sus cabezadas de la acequia del dicho valle, limpias y ensanchadas cada uno por su cuenta y cabezada una tercia entendiéndose que ha de ser mas del ancho que hoy trae la dicha acequia...

...Otros mandaron que luebo que hayan a cabado de limpiar de dicha acequia, todos juntos vayan a limpiar la acequia dicha que resta la cual se ha de limpiar desde la grangera que llaman...hasta llegar a la presa y toma del agua principal lo cual se han obligado a hacer todos y cada uno con la gente que turvieren so pena..."

Los extractos anteriores afirman categoricamente los derechos que ya habian adquirido los labradores del Valle de Santiago, y a mayor abundamiento de pruebas, un siglo mas tarde de la narración anterior el 25 de febrero de 1722 el Virrey don Baltazar de Zúñiga, Marqués de Valero expidió un despacho para que las justicias a quienes fuere presentado, midieran el agua del Río Grande del Valle de Santiago, arreglándose a la respuesta del asesor inserto, y ejecuten lo demás que se les previene.

El anterior despacho fue expedido por el Virrey, en virtud de una queja que le presenteron los labradores del Valle de Santiago por robo de aguas y cuya exposición dicen: ..."En el Río Grande de dicho Valle de Santiago tiene mi parte una presa general para que rebalse el agua y pueda salir de una acequia o conducto para conducirla a todas las haciendas de dicho Valle ..."y mas adelante dice..."en las expensas de la presa son todos iguales..."

El mismo Virrey Marqués de Valero expidió y otorgó la confirmación de un convenio que tuvieron los labradores en una junta el dia 4 de octubre de 1721, ante don Antonio Moreno de la Cerda, Alcalde Ordinario de Salamanca...

***

Con objecto de los aprovechamientos antiores, los labradores construyeron en la Presa de Santa Rita que primitivamente fue de madera y tierra pilares de calicanto, para proveerse de agua para sus terrenos en los meses de octubre y noviembre a enero o febrero epoca en que se acaba la del río, para volver a tomar la de mayo en adelante en que principian las corrientes del año.

Está, pues, probado que desde tiempos remotisimos los labradores del Valle han hecho uso de esas aguas para el tiempo de secas en los citados meses, y no podría presumirse de otra manera, toda vez que el albacenamiento para la laguna de Yuriria vino posteriormente, aliviando asi las necesidades que en años escasos tenían para sus labores, por no serles suficientes las corrientes del río, inclusivo sus "remanzos y manantiales"; por consiguiente, facil es comprender el derecho con que han cerrado y cierran la susodicha presa de Santa Rita en los terminos de tiempo señalados; y si en alguna epoca hubo un convenio para no cerrarla sino hasta el 15 de octubre, fué debido a la irregularidad de su sistema viejo. Dicho convenio no tuvo la solemnidad necesaria, y ha sido conceptuado como trasitorio y temporal con la hacienda de San Nicolás de los Agustinos, y se trató de conciliar interinamente los interese de ambas partes: pero desde que se establecio el sistema moderno de la presa (con compuertas de hierro), importantísima mejora verificada en 1910, quedó insubsistente el arreglo habido. Hoy se abren o se cierran dichas compuertas, en brevísimo tiempo, sin peligro para los terrenos contiguos.

***

Respecto a la concessión hecha a la hacienda de San José la Grande para disfrutar agua permanentemente por un orificio de una tercia de vara castellan (tercia en cuadro) quien estas memorias escribe tuvo a la vista un documento antiguo en pergamino en el que estaba consignado que en 1735 por convenio solemno celebrado entre la señora, doña María Peinado, Condesa de Rábago y don Pedro Bautista de Lascuráin y Retana, se obligó la primera a verificar la apertura de un canal desde su hacienda de La Grande hasta la de Guantes propiedad de los Padres Agustinos, porque el existente entonces solo llegaba desde la presa de Labradores situada abajo del pueblo del Jaral, hasta dicha finca de José la Grande, y por ello convinieron todos los labradores inclusivo la Sagrada Provincia en otorgarle la referida merced. En dicho documento en la parte expositiva esta literal de siguiente fragmento: ..."por cuanto al beneficio que reciba el padron del comun de labradores se le concede a San José la Grande una tercia de agua que fijara en su cabezada sin moverla de alli donde diga el perito y perenne en tiempo de tandas..."

El canal desde la laguna de Yuriria hasta la Presa de Labradores se hizo aprovechando el cauce viejo por donde algunas veces derramo sus aguas dicha laguna, cuando por la abundancia del temporal se llenaba demasiado; siguiendo un rumbo SSE a NNW pasando por el Pitayo en donde movia unos rodesnos para moler trigo, cayendo al rio.

De la Citada Presa de Labradores partía otro canal por donde primitivamente se regaban algunos terrenos. Hoy sirve como desagüe y se llama acequia vieja.

***

Posteriormente y como complemento de la anterior excritura de compromiso, se otrogo otra el 20 de diciembre de 1797 ante don José Carlos Valle, Regidor Fiel Ejecutor y alcalde ordinario, por autorizacion de D. Juan Antonio Riaño, Intendente de la provincia, fijando la verificación de obras de mampostería, amplicción de canales y otras cosas similares, entre ellas senalando el punto para el canal de captación de las aguas del río, y el de extracción en el sitio llamado El Remanso; conviniendo en que las aguas se habian de extraer en determinado tiempo y para determinados fines, cuando a la hacienda de la Bolsa, una tercia en cuadro para sacar agua permamentemente segun se estipulo en la escritura de compromiso, y de conformidad con el parage fijado para ello en la escritura de 20 de diciembre de 1797, o sea en el portillo de corrales sin variar la de dicho lugar, asi mismos e pacto en esa escritura, formar un bordo desde la puente de Taramatacheo (literal) para el oriente en todos los parages que lo pidiera la Sagrada Provincia para obviar que se extendiera el agua inutilizando terrenos de San Nicolás.

Glossary

acequia	irrigation ditch or canal.
alguacil	constable.
alcalde mayor	Spanish magistrate in charge of the highest level of local government, the alcaldía mayor.
alcaldía mayor	a unit of local government comparable to the English county.
alcalde ordinario	district magistrate.
audiencia	the highest court in the colonial judicial system.
barrio	Indian neighborhood in a community.
caballería	a unit of land equal to 105.76 acres or 42.8 hectares.
cabildo	municipal council.
cacique	a hereditary Indian chief.
carga	a unit of volume usually taken to equal two fanegas, but the tithe administrators of Michoacán used an equivalence of three fanegas per carga.
ciudad	a city; a designation applied only to towns of the highest rank.
congregación	a municipal unit; generally refers to a small settlement.
ejido	communal grazing lands of a municipality.
encomendero	holder of an encomiendo.
encomienda	a private right to receive Indian tribute, theoretically involving reciprocal obligations.
estancia	grazing land; a cattle, sheep, or goat ranch. Also refers to early viceregal grants of grazing rights from which many haciendas derived their title.
fanega	a unit of dry volume. The common Castilian fanega equaled 1.5 English bushels, but there is evidence that a larger measure, the Mexican fanega equal to about 2.5 bushels, may have been used in the Bajío.[1]
fiscal	Crown attorney in the audiencia.
hacendado	owner of a hacienda.
hacienda	a large rural estate with certain characteristic features that dominated the rural society of late colonial Mexico.
jagüey	pond used for irrigation or watering livestock.
labrador	a farmer of Spanish descent.
mayorazgo	an entailed estate that could not be alienated and that descended to the eldest son.

[1]Brading (1978), p. 66.

merced	a grant of land, water, or other privilege.
milpa	a plot of land; a cornfield.
noria	a waterlift.
oidor	a judge of the audiencia.
peón	a manual laborer.
peso	a silver coin constituting the basic monetary unit of colonial Mexico. Equaled one ounce of silver.
presa	a dam.
ranchero	the owner or renter of a rancho.
rancho	a small rural property, often on land rented from a hacienda.
real	one-eighth of a peso.
repartimiento	an allocation or distribution. In this essay it has four applications: (1) an allocation of forced Indian labor; (2) a distribution of land to vecinos of a town; (3) the distribution of water at an irrigation box; (4) a judicial allocation of water rights.
riego	irrigation.
tanda	a term with various meanings, but usually referring to an irrigation turn.
vara	equals 33 English inches or .838 meters. One vara = 3 pies = 36 pulgadas.
vecindad	a more or less standard allotment of land given to Spanish vecinos following the establishment of a villa or ciudad.
vecinos	a citizen of a municipality, ordinarily a property owner.
villa	a municipality ranking on level below a ciudad.

Archival Sources

Archivo General de la Nación, Mexico. Volumes consulted in whole in part, listed by *ramo*:

Tierras, 65, 69, 95, 99, 111, 115, 118, 119, 154, 155, 185, 187, 192, 232, 237, 238, 253, 272, 274, 275, 294, 307, 311, 351, 353, 380, 383, 394, 417, 432, 435, 440, 450, 455, 465, 487, 495, 499, 514, 533, 534, 535, 544, 548, 550, 570, 580, 586, 602, 618, 621, 627, 629, 630, 637, 666, 673, 674, 675, 764, 787, 789, 822, 824, 866, 868, 879, 972, 988, 1020, 1041, 1061, 1107, 1113, 1114, 1124, 1166, 1168, 1169, 1170, 1247, 1333, 1362, 1390, 1810, 1816, 1888, 1970, 2055, 2071, 2072, 2073, 2074, 2367, 2628, 2643, 2648, 2691, 2701, 2705, 2716, 2738, 2747, 2760, 2765, 2767, 2782, 2785, 2791, 2809, 2816, 2899, 2901, 2935, 2951, 2952, 2959, 2963, 2967, 2970, 3401, 3618, and 3687.

Mercedes, 7-14, 16, 18, 19, 21-26, 28-39, 41, 44, 45, 49, 50, 52, 53, 56, 58, 60, 70-72, 74, 75, 79, and 81.

Ayuntamientos, 97, 108, 138, 147, 169, 180, 218, and 225.

Civil, 73, 120, 164, 272, 359, and 1298.

General de Parte, 1, 2, and 4-6.

Historia, 72, 439, and 578B.

Padrones, 23, 26, and 45.

Industria y Comercio, 14 and 32.

Fomento Caminos, 40.

Indios, 3.

Reales Cédulas, 18.

Templos y Conventos, 24.

Vínculos, 166.

Archivo Casa Morelos, Morelia. Selected documents in the following legajos of the ramos of Capellanías and Diezmatorios:

legajos 815, 821, 823, 825, 830-38, 841, 847, 849, 852, 860, 861, and 862.

Museo Nacional de Antropología, Mexico.

Archivo Histórico:

Collección General, leg. 3698, doc. 8.

Eulalia Guzmán, leg. 104, doc. 11; 107, doc. 7.

Gómez de Orozco, leg. 1.

Collección Paso y Troncoso, leg. 33.

Fondo Franciscano, Vol. 92.

Ant. T-2, 1.

Archivo de Micropelículas (selected documents):
Archivo Judicial de Querétaro, rolls 5, 23, 27, 34, and 51.
Archivo Casa Morelos, rolls 776636 and 776638.

Archivo Franciscano de Celaya, selected documents, listed by section of archive:
Convento de Santa Clara, legajos "títulos" and "pleitos."
Archivo Provincial, legajo P-1.
Convento de Celaya, legajos C-3, D, E-2, E-4, L-4, and T-2.

Archivo Histórico de León. Selected documents in the following boxes:
Cajas, 1606-09, 1613-14, 1617-19, 1620, 1646-48, 1649-52, 1660-63, 1664-72, and 1685-90.

Archivo Municipal de Salvatierra.
Four documents classified under years 1652, 1707, and 1717.

Augustinian Archive, Mexico City.
Lib. Prov. Vol. 5
Two unclassified documents.

Bancroft Library, University of California, Berkeley.
Manuscripts of Sáenz de Escobar
PE-51:2 (El Paso)

Private archive of Basilio Rojas, Valle de Santiago.
Two documents.

Archivo General de Indias, Seville.
Indiferente, leg. 107, tomo 1, f. 386.

Biblioteca Nacional de México.
Fondo Franciscano, caja 47, doc. 1059.

Cathedral Archive of Archbishopric of Mexico.
Book 374 (tithe records).

Condumex, Mexico City.
Adquisiciones Diversas, fondo 507.

New York Public Library.
Manuscript: *Forma muy util y provechosa para medir aguas, sitios y caballerias de tierras...*

University of Texas, Austin, Latin American Documents Collection.
WSB 463.

Archivo de Secretaría de Reforma Agraria, Irapuato.
In several days of research, I did not find a single colonial document in the regional archive of the SRA, but the many early twentieth century maps were invaluable in identifying the location of haciendas. Without these maps, I could not have interpreted other documents.

As the above tabulation indicates, this essay has been based largely on sources in the Archivo General de la Nación, particularly the ramos of Tierras and Mercedes. Colonial archives have been lost or destroyed in the municipalities of Celaya, Apaseo, Valle de Santiago, Salamanca, Acámbaro, Yuriria, Irapuato, and Silao. Part of the archive of Salvatierra still survives but is in an advanced state of deterioration. I was able to locate four relevant documents only with the help of the chronicler of Salvatierra, Vicente Ruiz Arias. The Archivo Histórico Municipal of Guanajuato contains a scattering of documents on the nearby areas of Irapuato, Silao, and León, but very few on the areas that I studied in detail.

In historical research, one is seldom able to exhaust all possible sources, but work that might have been done should still be mentioned. First, it is regrettable that this study could not include a detailed reconstruction of the waterworks on the Rivers Silao and Guanajuato. The intensive exploitation of these rivers in the eighteenth century was the product of social forces significantly different from those that affected the irrigation systems of Salvatierra, Valle de Santiago, Celaya, and Querétaro. Unfortunately, the documentary resources on Irapuato and Silao in the Archivo General de la Nación are very meagre; and a search of the most promising local archives, the Archivo Histórico Municipal of Guanajuato and the Archivo Casa Morelos, would have required much more time than I could afford and might well have proved to be unproductive. Secondly, it would have been desirable to research more thoroughly the very large Augustinian Archive, housed in the church of San Agustín in the Polanco district of Mexico City. Archbishop John R. Quinn, then president of the U.S. Conference of Catholic Bishops, generously wrote a letter to the Augustinian order recommending that I be given access to this archive. When shown the letter, the Augustinian archivist extracted a small selection of documents from the archive that I was allowed to read rapidly, under close watch, for a period of about five hours. The archivist refused me any further access to the archive with the explanation that it contained "secret documents." Thirdly, a truly exhaustive study of Querétaro would have involved research of the judicial archive itself rather than merely the selected documents from the archive in the microfilm collection of the Museo Nacional de Antropología. Because of limitations of time, I contented myself with reading relevant documents in the microfilm collection that were mentioned in the guide to the collection or in John Super's book on Querétaro (1982).

Bibliography

Manuscript Treatises

Forma muy util y provechosa para medir aguas, sitios y cavallerias de tierras de ganados maiores y menores y la practica judicial que deben observar las justicias. n.d. New York Public Library.

Sáenz de Escobar, José. *Ordenanzas vigentes para las medidas de tierras y aguas de este America Septentrional.* n.d. Bancroft Library, University of California.

__________. *Libro de las ordenanzas y medidas de tierras y aguas.* n.d. Bancroft Library; and Biblioteca Nacional, Mexico.

Santa Cruz Talaban, Alejandro de la. 1778. *Elementos de pintura, statica, hydraulica y algebra.* Vol. 4. Museo Nacional de Antropología, Mexico.

Published Sources

Altamira, Rafael. 1902. Mercado de agua para riego en la huerta de Alicante y en otras localidades de la Peninsula. In *Derecho consuetudinario y economía popular de España,* Joaquín Costa, et al., Vol. 2, pp. 136-64, 441-47. Barcelona: Manual Solar.

Alzate y Ramírez, José Antonio. 1831. *Gacetas de literatura de México.* 4 Vols. Puebla.

__________. 1768. *Diario literario de México.* Mexico: Imp. de la Bibliotheca Méxicana.

Arneson, Edwin P. 1921. Early irrigation in Texas. *Southwestern Historical Quarterly* 25:120-30.

Arregui, Domingo Lázaro de. 1946. *Descripción de la Nueva Galicia,* ed., Francois Chevalier, No. 24, Series 3. Sevilla: Publicaciones de la escuela de Estudios Hispano-Americanos de la Universidad de Sevilla.

Asana, R.D., and Saini, A.D. 1962. Studies in physiological analysis of yield. *Indian Journal of Plant Physiology* 5:128-72.

Asana, R.D., and Williams, R.F. 1965. The effect of temperature stress on grain development in wheat. *Australian Journal of Agricultural Research* 16:1-13.

Báez Macías, Eduardo. 1965. Tres mapas de los siglos XVII y XVIII sobre la ciudad de Salvatierra. *Boletín del Archivo General de la Nación* 6:671-711.

Bakewell, P.J. 1971. *Silver Mining and Society in Colonial Mexico: Zacatecas 1546-1700.* Cambridge, UK: University of Cambridge Press.

Bancroft, Hubert Howe. 1883-88. *History of Mexico.* 6 Vols. San Francisco: A.L. Bancroft.

Basalenque, Diego. 1963. *Historia de la provincia de San Nicolás de Tolentino de Michoacán del orden de N.P.S. Agustín*. Mexico: Editorial Jus.

Basave Kunhardt, Jorge. 1977. Algunos aspectos de la técnica en los haciendas. In *Siete Ensayos sobre la hacienda*, ed. Enrique Semo, pp. 189-246. Mexico: Departmento de Investigaciones Históricas, INAH.

Bazant, Jan. 1975. *Cinco haciendas Méxicanas, Tres Siglos de Vida Rural en San Luis Potosí*. Mexico: El Colegio de Mexico.

Beardsley, Richard K. 1964. Ecological and social parallels between rice growing communities of Japan and Spain. In *Symposium on Community Studies in Anthropology*, ed. V.E. Garfield, pp. 51-63. Proceedings of the 1963 Annual Spring Meeting of the American Ethnological Society. Seattle: American Ethnological Society.

Beaumont, Pablo. 1874. *Crónica de la provincia de los Santos Apóstoles San Pedro y San Pablo de Michoacán*. Mexico: Imprenta de Ignacio Escalante.

Beleña, Eusebio Ventura. 1787. *Recopilación sumaria de todos los autos acordados de la real audiencia y sala del crimen de esta Nueva España y providencias de su superior gobierno*. 2 Vols. Mexico: Zuñiga y Ontiveros.

Borah, Woodrow, and Cook, Sherburne F. 1958. *Price Trends of Some Basic Commodities in Central Mexico, 1531-1570*. Berkeley and Los Angeles: University of California Press.

Borde, Jean, and Góngora, Mario. 1965. *Evolución de la propiedad rural en el valle de Puangue*. Santiago de Chile: Editorial Universitario.

Brading, David A. 1971. *Miners and Merchants in Borbon Mexico, 1763-1810*. Cambridge, UK: Cambridge University Press.

__________. 1973. La estructura de la producción agrícola en el Bajío de 1700 a 1850. *Historia Méxicana* 23:197-237.

__________. 1978. *Haciendas and Ranchos in the Mexican Bajio, 1700-1860*. Cambridge, UK: Cambridge University Press.

Brading, David A., and Wu, Celia. 1973. Population growth and crisis: León 1720-1860. *Journal of Latin American Studies* 5:1-36.

Bribiesca Castrejón, José Luis. 1958-59. Agua potable--historia en Mexico. *Ingeniería hidráulica en México*. Vols. 12(2) to 13(1).

Bullock, William. 1824. *Six Months Residence and Travel in Mexico*. London: John Murray.

Cajori, Florian. 1928. *The Early Mathematical Sciences in North and South America*. Boston: Gordon Press.

Canvin, David T., and Yao, Yun-te. 1967. Effect of temperatures on the growth of wheat. *Canadian Journal of Botany* 45:757-72.

Carrera Stampa, Manuel. 1967. El sistema de pesos y medidas coloniales. *Memorias de la Academia Méxicana de la Historia* 26:1-37.

Cartografía de Querétaro. 1978. Querétaro: Gobierno del Estado.

Castorena y Ursúa, Juan Ignacio and Sahagún de Arevalo, Juan Francisco, eds. 1949. *Gacetas de Mexico*. 3 Vols. Mexico: Secretaría de Educación Pública.

Cavo, Andrés. 1852. *Los tres siglos de México durante el Gobierno Español*. Mexico: J.R. Navarro.

Chávez Orozco, Luis. 1950. La irrigación en México, ensayo histórico. *Problemas agrícolas e industriales de México* 2:11-32.

Chester, K. Starr. 1946. *Nature and Prevention of Cereal Rusts*. Waltham, MA: Chronica Botanica Company.

Chevalier, Francois. 1952. *La Formation des Grands Domaines au Mexique*. Travaux y Memoires de Institut d'Ethnologie, No. 56. Paris: Institut d'Ethnologie.

Chevalier, Francois, ed. 1950. *Instrucciones a los hermanos jesuitas administradores de haciendas*. Mexico: Instituto de Historia, UNAM.

Chinchilla Aguilar, Ernesto. 1953. El ramo de aguas de la ciudad de Guatemala en la época colonial. *Antropología e historia de Guatemala* 5:19-31.
Cook, Sherburne F., and Borah, Woodrow. 1971-79. *Essays on Population History: Mexico and the Caribbean*. 3 Vols. Berkeley: University of California Press.
Corona Núñez, José, ed. 1958. *Relaciones geográficas de la diócesis de Michoacán--1579-80*. Collección Siglo XVI, 2 Vols. Guadalajara.
Cuevas Aguirre y Espinosa, José Francisco. 1748. *Extracto de los autos de diligencias y reconocimientos de los ríos, lagunas, vertientes y desagües de la capital de México y su valle*. Mexico: Vda. de D. Joseph Bernardo de Hogal.
Daugherty, Robert C., and Franzini, Joseph B. 1977. *Fluid Mechanics with Engineering Applications*. New York: McGraw Hill.
Daumas, Maurice. 1953. *Les Instruments Scientifiques aux XVIIe et XVIIIe Siècles*. Paris: Presses Universitaires de France.
Diario de México, 1810. 1810. Mexico: Zuñiga y Ontiveros.
Dobkins, Betty Eakle. 1959. *The Spanish Element in Texas Water Law*. Austin: University of Texas Press.
Downing, Theodore E., and Gibson, McGuire, eds. 1974. *Irrigation's Impact on Society*. Anthropological Papers of the University of Arizona, No. 25. Tucson: University of Arizona Press.
Dunne, Thomas, and Leopold, Luna B. 1978. *Water in Environmental Planning*. San Francisco: W.H. Freeman.
Escobar, Matías de. 1970. *Americana thebaida: vitas patrum de los religiosas hermitanos de N.P. San Agustín de la provincia de San Nicolás de Michoacán*. Morelia: Balsal Editores.
Escriche y Martín, Joaquín. 1838. *Diccionario razonado de legislación civil, penal, comercial y forense*. 2d ed. 3 Vols. Madrid: Imprenta del Colegio Nacional de Sordomudas.
Espinosa, Isidro Félix de. 1945. *Crónica de la provincia franciscana de los apostoles San Pedro y San Pablo de Michoacán*. 2d ed. Mexico: Editorial Santiago.
Estado general de las fundaciones hechas por D. José de Escandón en la colonia del Nueva Santander. 1929. Mexico: Talleres Gráficos de la Nación.
Ewald, Ursula. 1976. *Estudios sobre la hacienda colonial en México: las propiedades rurales del Colegio Espiritú Santo en Puebla*. Wiesbaden, FRG: Franz Steiner Verlag.
Fernández del Castillo, Francisco. 1927. *Tres conquistadores y pobladores de la Nueva España*. Mexico: Talleres Gráficas de la Nación.
Fernández y Fernández, Ramón. 1935. Historia del trigo. *Trimestre económico* 1:429-44.
Fernández de Recas, Guillermo. 1965. *Mayorazgos de la Nueva España*. Mexico: Instituto Bibliográfico Méxicano.
Fernández y Simón, Abel. 1950. *Memoria histórica-técnica de los acueductos de la Habana*. Primera Parte. La Habana: Impresores Ucar García.
Florescano, Enrique. 1969. *Precios del maíz y crisis agrícolas en México*. Mexico: El Colegio de México.
__________. 1976. *Origen y desarrollo de los problemas agrarios de México: 1500-1820*. Mexico: Ediciones Era.
Florescano, Enrique, and Sánchez Gil, Isabel, eds. 1973. *Descripciones económicas generales de Nueva España, 1784-1817*. Mexico: Departamento de Investigaciones Históricas, INAH.
Frank, André Gunder. 1979. *Mexican Agriculture 1521-1630: Transformation of the Mode of Production*. Cambridge, UK: Cambridge University Press.
Friend, D.J.C. 1965. Tillering and leaf production in wheat as affected by temperature and light intensity. *Canadian Journal of Botany* 43:1066-76.

Galicia Morales, Silvia. 1975. *Precios y producción en San Miguel el Grande, 1661-1803*. Mexico: Departamento de Investigaciones Históricas, INAH.
Galván Rivera, Mariano. 1842. *Ordenanzas de tierras y aguas o sea formulario geométrico-judicial*. Mexico: Imp. de Vicente Torres.
Gamboa, Francisco Javier de. 1761. *Comentarios a las ordenanzas de minas*. Madrid: Joachim Ibarra.
García Miranda, Alfonso. 1940. La concentración del suelo agrícola. *El Trimestre Económico* 7:494-513.
Geertz, Clifford. 1972. The wet and the dry: traditional irrigation in Bali and Morocco. *Human Ecology* 1:23-39.
Gemelli Careri, Juan Francisco. 1927. *Viaje a la Nueva España*. 2 Vols. Mexico: Sociedad de Bibliófilos Méxicanos.
Gerhard, Peter. 1962. *México en 1742*. Mexico: José Porrúa e Hijo.
__________ 1972. *A Guide to the Historical Geography of New Spain*. Cambridge Latin American Studies, No. 14. Cambridge, UK: Cambridge University Press.
Gibson, Charles. 1964. *The Aztecs under Spanish Rule, 1519-1810*. Stanford: Stanford University Press.
Gillispie, Charles Couston, ed. 1975. *Dictionary of Scientific Biography*. 16 Vols. New York: Charles Scribner.
Glick, Thomas F. 1968. Levels and levelers: surveying irrigation canals in medieval Valencia. *Technology and Culture* 9:165-80.
__________. 1972. *The Old World Background of the Irrigation Systems of San Antonio, Texas*. Southwestern Studies Monograph, No. 35. El Paso: Texas Western Press of the University of Texas at El Paso.
Gómez Pérez, Francisco. 1942. Mexican irrigation in the 16th century. *Civil Engineering* 12:24-27.
Gonzales, Pedro. 1904. *Geografía local del estado de Guanajuato*. Guanajuato: Tip. de la Escuela Ind. Militar.
Gray, Robert F. 1974. *The Sonjo of Tanganyika*. Westport, CT: Greenwood Press.
Humboldt, Alexander von. 1811. *Essai Politique sur le royaume de la Nouvelle Espagne*. 4 Vols. Paris: F. Schoell.
Hunt, Robert C., and Hunt, Eva. 1976. Canal irrigation and local social organization. *Current Anthropology* 17:388-411.
Hurtado López, Flor de María. 1974. *Dolores Hidalgo: estudio económico, 1740-1790*. Mexico: Departamento de Investigaciones Históricas, INAH.
Hutchins, Wells A. 1927. The community Acequia: its origin and development. *Southwestern Historical Quarterly* 31:261-84.
Informacion de los méritos y servicios prestados por don Fernando de Tapia en la conquista y fundación de Querétaro y provanza del cacicazgo de don Diego de Tapia. *Boletín del Archivo General de la Nación* 5(1934):34-61.
Jáuregui, Ernesto, Klaus, D., and Lauer, W. 1977. Estimación de la evaporación y evapotranspiración potencial del centro de México. *Recursos Hidráulicos* 6:11-25.
Jiménez Jaime, Enrique. 1978. *Crónica de Celaya*. Celaya: Impresora Gutenberg.
Jiménez Moreno, Wigberto. 1956. *Estudios de historia colonial*. Mexico: Instituto Nacional de Antropología e Historia.
Johnson, Frederick, ed. 1972. *The Prehistory of the Tehuacan Valley*. Austin: University of Texas Press.
Kakde, J.R. 1963. Some considerations of lodging in wheat. *Poona Agricultural College Magazine* 54:37-41.
Kiely, Edmond R. 1947. *Surveying Instruments: Their History and Classroom Use*. New York: Teachers College, Columbia University.
Keith, Robert C. 1976. *Conquest and Agrarian Change: the Emergence of the Hacienda System on the Peruvian Coast*. Cambridge, MA: Harvard University Press.

Konetzke, Richard, ed. 1953-62. *Colección de documentos para la historia de la formación social de Hispano-américa, 1493-1810*. 3 Vols. in 5 parts. Madrid: Concejo Superior de Investigaciones Científicas.
La Hire, Philippe. 1689. *L'Ecole des Arpenteurs*. Paris: Thomas Moette.
Lasso de la Vega, Domingo. 1761. *Reglamento general de las medidas de las aguas dispuesto a la municipal de esta governación*. Mexico: Imp. de la Bibliotheca Méxicana.
Lavrin, Asunción. 1975. El Convento de Santa Clara de Querétaro--la administración de sus propiedades en el siglo XVII. *Historia Méxicana* 25:76-118.
Lee, Raymond L. 1947. Grain legislation in colonial Mexico. *Hispanic American Historical Review* 27:647-60.
León, Nicolás de. 1903. *Bibliografía Méxicana del siglo XVIII*. Vols. 1-5. Mexico: J.I. Guerrera y Cia.
Lira, Andrés, and Muro, Luis. 1976. El siglo de la integración. In *Historia general de México*, pp. 83-182. Vol. 2. Mexico: El Colegio de México.
Llauradó, Andrés. 1884. *Tratado de aguas y riegos*. 2 Vols. Madrid: Moreno y Rojas.
Loegering, W.Q., Johnston, C.O., and Hendrix, J.W. 1969. Wheat Rusts. In *Wheat and Wheat Improvement*, ed. K.S. Quisenberry, pp. 307-36. Agronomy Series No. 13. Madison, WI: American Society of Agronomy.
López Sarrelangue, Delfina E. 1966. Las tierras comunales indígenas en la Nueva España en el siglo XVI. *Estudios de historia novo-hispana* 1:131-49.
Luquín, Carlos, Rojas Garcidueñas, José, and de la Maza, Francisco. 1941. *Apuntes para la historia de los aprovechamientos hidráulicos en México*. Mexico: Comisión Nacional de Irrigación.
Malagón Barcelo, Javier. 1959. *La literatura jurídica Española del siglo de oro en la Nueva España*. Mexico: Biblioteca Nacional de Mexico, Instituto Bibliográfico Méxicano.
Marco Cuellar, Roberto. 1965. El compendio mathematico del P. Tosca y la introducción de la ciencia moderna. *Actas II Congreso Español de la historia de medicina* 1:325-58.
Mariátegui, José Carlos. n.d. *Siete ensayos de interpretación de la realidad Peruana*. Lima: Distribuidora Jucen.
Martín, Enrico. 1948. *Repertorio de los tiempos e historia natural de Nueva España*. Mexico: Secretaría de Educación.
Martínez Luna, Victor Manuel. 1980. *Los factores geomorfológicos que rigen el comportamiento de la presa "Ignacio Allende", Guanajuato*. Mexico: Instituto de Geografía, UNAM.
Martínez Ríos, Jorge. 1970. *Tenencia de la tierra y desarrollo agraria en México: bibliografía selectiva y comentada: 1522 a 1968*. Mexico: Instituto de Investigaciones Sociales, UNAM.
Mas Sinta, Juan. 1969. *Condiciones climatológicas predominantes del estado de Guanajuato*. Mexico: Sociedad de Geografía y Estadística.
Medina, José Toribio. 1907. *La imprenta en México, 1539-1821*. 8 Vols. Santiago de Chile: casa del autor.
Mexico. 1978. *Ley federal de aguas*. 5th ed. Mexico: Editorial Porrúa.
Mexico, Secretaría de Agricultura y Ganadería. 1956. *Estudios que está realizando el Instituto de Investigaciones Agrícolas de México para obtener mayor resistencia a los diferentes razas de chahuiztle*. Mexico: Instituto de Investigaciones Agrícolas.
Mexico, Secretaría de Recursos Hidráulicos. 1970. *Boletín hidráulico, no. 51, región hidrológica 12*. Mexico.
__________. 1973. *Estudio geomorfológico de la cuenca del Río de la Laja*. Coordinated by Rubén López Recéndez. Mexico.

__________. 1975. *Estudio geohidrológico cuantitativo de los acuíferos del Alto Lerma, Guanajuato*. Mexico.
__________. 1976. *Atlas de agua*. Mexico.
Miranda, José. 1962. *Humboldt y Mexico*. Mexico: Instituto de Historia, UNAM.
Moreno, Roberto. 1969. Catálogo de los manuscritos científicos de la biblioteca nacional. *Boletín del Instituto de Investigaciones Bibliográficas* 1:61-104.
__________. 1977. *Joaquín Velázquez de León y sus trabajos científicos sobre el valle de México, 1773-1775*. Mexico: Instituto de Investigaciones Históricas, UNAM.
Moreno Toscano, Alejandra. 1974. Economía regional y urbanización: tres ejemplos de relación entre ciudades y regiones en Nueva España a finales del siglo XVIII. In *Ensayos sobre el desarrollo urbano de México*, eds. Edward E. Calneck, et al., pp. 95-130. Mexico: SepSetentas.
Morfí, Juan Agustín de. 1967. *Diario y derrotero, 1777-1781*. Monterrey: Instituto Técnico y de Estudios Superiores de Monterrey.
Morin, Claude. 1979. *Michoacán en la Nueva España del siglo XVIII, crecimiento y desigualdad en una economía colonial*. Mexico: Fondo de Cultura Económica.
Mota y Escobar, Alonso de la. 1966. *Descripción geográfica de los reinos de Nueva Galicia, Nueva Viscaya y Nueva León*. Guadalajara: Instituto Jalisciense de Antropología e Historia.
Mumford, Lewis. 1961. *The City in History*. New York: Harcourt, Brace & World.
Navarrete, Francisco Antonio. 1961. *Relación peregrina del agua de Querétaro*. Mexico: Bibliófilos Méxicanos.
Navarrete, Nicolás. 1978. *Historia de la provincia Agustiniana de San Nicolás de Tolentino de Michoacán*. 2 Vols. Mexico: Biblioteca Porrúa.
Novísima recopilación de las leyes de España. 1805-1807. 6 Vols. Madrid.
Nueva recopilación (see Pérez y López).
El obispado de Michoacán en el siglo XVII, informe inédito de beneficios, pueblos y lenguas. 1973. Morelia: Fimas Publicistas.
Olson, Lois, and Eddy, Helen L. 1943. Ibn-al-Alwam: a soil scientist of Moorish Spain. *Geographical Review* 33:100-109.
Orozco y Berra, Manuel. 1854. Medidas y pesos en la república Méxicana. In *Diccionario universal de historia y geografía*, pp. 206-14. Vol. 5. Mexico: Imp. de F. Escalante y Cia.
Ostrom, Vincent. 1953. *Water and Politics: a Study of Water Policies and Administration in the Development of Los Angeles*. Los Angeles: Haynes Foundation.
Penman, H.L. 1956. Estimating evaporation. *Transactions, American Geophysical Union* 37:43-50.
Pérez y López, Antonio Xavier. 1791. *Teatro de la legislación universal de España e Indias por orden cronológico de sus cuerpos y decisiones no recopiladas*. 28 Vols. Madrid: Imp. de Manual Gonzales.
Piña Dávila, Margarita, Reyes Chirinos, Cihuapilli Virginia, and Bonilla Burgos, Luz María. 1980. *Geografía de la región de Valle de Santiago, Guanajuato*. Thesis, Escuela Normal Superior de México, México, D.F..
[Ponce, Alonso]. 1873. *Relacion breve y verdadera de algunas cosas de las muchas que sucedieron al Padre Fray Alonza Ponce en los provincias de Nueva España*. 3 Vols. Madrid: Imp. de la vda. de Calero.
Powell, Philip Wayne. 1952. *Soldiers, Indians and Silver, the Northward Advance of New Spain, 1550-1600*. Berkeley and Los Angeles: University of California Press.
Prem, Hanns J. 1974. El Río Cotzala: estudio histórico de un sistema de riego. *Comunicaciones* 11:53-67.
Price, Barbara J. 1971. Prehispanic irrigation in nuclear America. *Latin American Research Review* 6:3-60.

Primeros ordenanzas de la muy noble y muy leal ciudad de Querétaro. 1956. Querétaro: Editorial Querétaro.

Ramírez, M., Guillermina. 1976. *El Marqués del Villar del Aguila y la construcción del acueducto de Querétaro*. Thesis, Universidad Iberoamericana, Mexico D.F.

Raso, José Antonio. 1848. *Notas estadísticas del departamento de Querétaro formados por la asamblea constitucional del mismo y renditos al supremo gobierno en cumplimiento de la parte primera del articulo 135 de las bases orgánicas, año de 1845*. Mexico: Imprenta de José Lara.

Rea, Alonso de la. 1882. *Crónica de la orden de N. Seráfico P. San Francisco provincia de San Pedro y San Pablo de Mechoacán en la Nueva España*. Mexico: Imp. de J.R. Barbedilla.

Recopilación de leyes de los reynos de las Indias. 1791. 30 Vols. Madrid: Vda de J. Ibarra.

Relaciones del desagüe del valle de México. 1976. Mexico: Secretaría de Obras Públicas.

Ressler, John Q. 1968. Indian and Spanish water control on New Spain's northwestern frontier. *Journal of the West* 7:10-17.

Richeson, A.W. 1966. *English Land Measures to 1800: Instruments and Practices*. Cambridge, MA: MIT Press.

Rojas, Basilio. 1971. Breve síntesis histórica de Valle de Santiago. Conferencia dictada a la Sociedad de Geografía y Estadística de Guanajuato, July 7.

__________. 1976. El regadío del Bajío. In *Resumen informativo de la dirección de construcción de la dirección general de irrigación y control de ríos, Secretaría de Recursos Hidráulicos*, No. 17. México, D.F.: Secretaría de Recursos Hidraúlicos.

Rojas, R., Teresa; Strauss K., Rafael; Lameiras, José. 1974. *Nuevas Noticias sobre las obras hidráulicas prehispánicas y coloniales en el valle de México*. Mexico: Centro de Investigaciones Superiories, INAH.

Rodríguez Frausto, Jesús. 1976. *León se fundó así*. Guanajuato: Archivo Histórico.

Rodríguez de San Miguel, Juan N. 1852. *Pandectas Hispano-Mégicanos*. 3 Vols. Mexico.

Romero, José Guadalupe. 1949. *Noticias históricas de los pueblos del estado de Guanajuato*. Mexico: Editor Vargas Rea.

Romero de Terrero, Manuel. 1949. *Los acueductos de México en la historia y en el arte*. Mexico: Instituto de Investigaciones Estéticas, UNAM.

Rouse, Hunter, and Ince, Simon. 1957. *History of Hydraulics*. Iowa: Institute of Hydraulic Research, State University of Iowa.

Ruiz Arias, Vicente. 1976. *Historia civil y eclesiástica de Salvatierra, primera parte*. Mexico: Archivo Histórico de Salvatierra.

__________. 1980. *San José del Carmen*. Mimeographed. Salvatierra.

Ruiz-Funes García, Mariano. 1916. *Derecho consuetudinario y economía popular de la provincia de Murcia*. Madrid: Jaime Ratés.

Samaniego, Federico de. 1946. *El Marquéz de la Villa del Villar del Aguila*. Querétaro: Edición Cimataria.

Sanders, William T., and Price, Barbara J. 1968. *Meso-America: the Evolution of a Civilization*. New York: Random House.

Sanders, William T., and Marino, Joseph. 1970. *New World Prehistory*. Englewood Cliffs: Prentice Hall.

Sandoval, Fernando B. 1951. *La industria del azúcar en Nueva España*. Mexico: Instituto de Historia, UNAM.

Santa Cruz F., Iris E., and Giménez-Cacho García Luis. 1977. Los pesos y medidas en la agricultura. In *Siete ensayos sobre la hacienda méxicana, 1780-1880*, ed. Enrique Semo, pp. 247-69. Mexico: Departamento de Investigaciones Históricas.

Schott, Gaspar. 1660. *Pantometrum Kircherianum, hoc est, Instrumentum Geometricum Novum a Celeberrimo Viro P. Athanasio Kirchero.* Frankfurt: Johan Godefried Schonwetter.
Septien y Septien, Manuel. 1974. *Historia del acueducto y fuentes de Querétaro.* Querétaro: Edicion Culturales del Estado.
Septien y Villaseñor, José Antonio. 1875. *Memoria estadística del estado de Querétaro precedida de una noticia histórica.* Querétaro: Tip. de Gonzales y Ligarneto.
Serrera Contreras, Ramón. 1973. La ciudad de Santiago de Querétaro a fines del siglo XVII: apuntes para su historia urbana. *Anuario de estudios Americanos* 30:489-555.
__________. 1977. *Guadalajara ganadera: estudio regional novohispano 1760-1805.* Sevilla: Escuela de Estudios Hispano-Americanos.
Sherbondy, Jeanette. 1982. *The Canal Systems of Cuzco.* Thesis, University of Illinois.
Las siete partidas del Sabio Rey D. Alfonso con la glosa. del Lic. Gregorio López. 1844. 2 Vols. Barcelona: Imprenta de Antonio Bernges.
Sigüenza y Góngora, Carlos de. 1945. *Glorías de Querétaro en la nueva congregación eclesiástica de María Santísima de Guadalupe.* Querétaro.
Simons, Marc. 1972. Spanish irrigation practices in New Mexico. *New Mexico Historical Review* 47:135-50.
Solórzano Pereira, Juan de. 1930. *Política indiana.* Madrid: Cia. Ibero-Americano de Publicaciones.
Spota, Alberto G. 1941. *Tratado de derecho de aguas.* Buenos Aires: Casa Editorial de Jesús Menéndez.
Stanislawski, Dan. 1947. Tarascan political geography. *American Anthropologist* 49:46-55.
Steward, Julian H., et al. 1955. *Irrigation Civilizations: a Comparative Study.* Washington, DC: Pan American Union.
Stone, Edward Noble. 1926. Roman surveying instruments. In *University of Washington Publications in Language and Literature*, pp. 215-42. Vol. 4. Seattle: University of Washington Press.
Super, John C. 1976. Querétaro obrajes: industry and society in provincial Mexico, 1600-1810. *Hispanic American Historical Review* 56:197-216.
__________. 1982. *Querétaro.* Mexico: Fondo de Cultura Económica.
Swan, Susan Linda. 1977. Climate Crops and Livestock: Some Aspects of Colonial Mexican Agriculture. Thesis, Washington State University.
Tablas para la instrucción de las medidas de tierras y aguas formadas con arreglo a las ordenanzas vigentes que se observan en los estados unidos Méxicanos. 1827. Mexico: Imp. de la Testamentaria de Ontiveras.
Tamayo, Jorge L. 1962. *Geografía general de México.* 4 Vols. Mexico: Instituto de Investigaciones Económicas.
Taylor, William B. 1972. *Landlord and Peasant in Colonial Oaxaca.* Stanford: Stanford University Press.
__________. 1975. Land and water rights in the viceroyalty of New Spain. *New Mexico Historical Review* 50:189-212.
Testamento de Doña Beatriz de Tapia hija del conquistador de Querétaro Don Fernando de Tapia. *Boletín del Archivo General de la Nación* 17(1946):473-90.
Thornthwaite, C.W. 1948. An approach toward a rational classification of climate. *The Geographical Review* 38:55-94.
Título y fundación de la villa de Salamanca, 1602. *Boletín del Archivo General de la Nación* 6(1935):713-21.
Toussaint, Manuel. 1948. *Arte colonial en México.* Mexico: Imp. Universitaria.

Tovar Pinzón, Hermes. 1975. Elementos constitutivos de la empresa agraria jesuita en la segunda mitad del siglo XVIII en México. In *Haciendas, latifundios y plantaciones en América Latina*, coordinated by Enrique Florescano, pp. 132-222. Mexico: Siglo XXI Editores.
Valdés, Manual Antonio, ed. 1787. *Gazetas de México, compendio de noticias de Nueva España desde principios del año de 1784*. 16 Vols. Mexico: Zuñiga y Ontiveras.
Vázquez de Espinosa, Antonio. 1944. *Descripción de la Nueva España en el siglo XVII*. Mexico: Editorial Patria S.A.
Velasco y Mendosa, Luis. 1967. *Historia de la ciudad de Celaya*. 4 Vols. Mexico: Imp. Manuel León Sánchez.
Velázquez, Primo Feliciano. 1897. *Colección de documentos para la historia de San Luis Potosí*. 4 Vols. San Luis Potosí.
Vera, Melchor. 1939. *Guatzindeo Salvatierra*. San Luis Potosí, México, D.F.: Tipografía Moderna de J. Nieto Morales.
Vera Quintana, R. 1869. Noticias estadísticas de la ciudad de Salvatierra. *Boletín de sociedad Méxicana de geografía y estadística*. Vol. 1, 2a epoca, pp. 579-94.
Villanueva U., Sherbondy, Horacio, and Sherbondy, Jeanette. 1978. *Cuzco: aguas y poder*. Cuzco, Peru: Centro de Estudios Rurales Andinos.
Villaseñor y Sánchez, José Antonio de. 1952. *Theatro americano descripción general de los reynos y provincias de la Nueva España y sus jurisdicciones*. 2 Vols. Mexico: Editorial Nacional.
Vivó Escoto, Jorge A. 1964. Weather and climate of Mexico and Central America. In *Handbook of Middle American Indians*, ed., Robert Wauchope, pp. 187-215. Vol. 1. Austin: University of Texas Press.
Waitz, Paul. 1943. Reseña geológica de la cuenca de Lerma. *Boletín de la sociedad Méxicana de geografía y estadística* 58:123-38.
Wallen, C.C. 1956. Fluctuations and variability in Mexican rainfall. In *The Future of Arid Lands*, ed., Gilbert F. White, pp. 144-55. Pub. 43. Washington, DC: American Association for the Advancement of Science.
Ward, H.G. 1828. *Mexico in 1827*. 2 Vols. London: Henry Colburn.
Wattal, P.N. 1964. Effect of temperature on the development of the wheat grain. *Indian Journal of Plant Physiology* 8:145-59.
Wauters, Carlos. 1931. Origen y significado de las antiguas medidas de aguas en el interior regado. *Revista de la Universidad de Buenos Aires* 5:39-76.
Wellhausen, E.J., Roberts, L.M., and Hernández X., E. 1955. *Races of Maize in Mexico*. Cambridge, MA: Bussey Institution of Harvard University.
Wiese, M.V. 1977. *Compendium of Wheat Diseases*. St. Paul: American Phytopathological Society.
Wittfogel, Karl A. 1957. *Oriental Despotism: a Comparative Study of Total Power*. New Haven, CT: Yale University Press.
Wolf, Eric. 1957. The Mexican Bajío in the 18th century: an analysis of cultural integration. In *Synoptic Studies of Mexican Culture*, eds. Munro S. Edmonson et al., pp. 177-99. Middle American Research Institute Publication, No. 17. New Orleans: Tulane University.
Zamarroni Arroyo, Rafael. 1959-60. *Narraciones y leyendas de Celaya y del Bajío*. 2 Vols. Mexico: Editorial Periodística e Impresora de Mexico.
Zavala, Silvio Arturo. 1939-46. *Fuentes para la historia del trabajo en Nueva España*. Mexico: Fondo de Cultura Económica.
__________. 1947. *Ordenanzas del trabajo-siglo XVI-XVII*. Mexico: Editorial Elede.
Zeláa e Hidalgo, Joseph María. 1803. *Glorias de Querétaro*. Mexico: Zuñiga y Ontiveros.

Index

Dellplain Latin American Studies
Published by Westview Press

No. 8 *Studies in Spanish American Population History,* edited by David J. Robinson

No. 9 *The People of the Colca Valley: A Population Study,* Noble David Cook

No. 10 *Tierra Adentro: Settlement and Society in Colonial Durango,* Michael M. Swann

No. 11 *Andean Reflections: Letters from Carl O. Sauer While on a South American Trip Under a Grant from the Rockefeller Foundation, 1942,* edited by Robert C. West

No. 12 *Credit and Socioeconomic Change in Colonial Mexico: Loans and Mortgages in Guadalajara, 1720–1820,* Linda Greenow

No. 13 *Once Beneath the Forest: Prehistoric Terracing in the Rió Bec Region of the Maya Lowlands,* B. L. Turner, II

No. 14 *Marriage and Fertility in Chile: Demographic Turning Points in the Petorca Valley, 1840–1976,* Robert McCaa

No. 15 *The Spatial Organization of New Land Settlement in Latin America,* Jacob O. Maos

No. 16 *The Anglo-Argentine Connection, 1900–1939,* Roger Gravil

No. 17 *Costa Rica: A Geographical Interpretation in Historical Perspective,* Carolyn Hall

No. 18 *Household Economy and Urban Development: São Paulo, 1765 to 1836,* Elizabeth Anne Kuznesof

No. 19 *Irrigation in the Bajío Region of Colonial Mexico,* Michael E. Murphy